CUTTING EDGE RESEARCH IN TECHNOLOGIES

Edited by **Constantin Volosencu**

Cutting Edge Research in Technologies
http://dx.doi.org/10.5772/59460
Edited by Constantin Volosencu

Contributors

Ivana Schwarz, Pinaki Mitra, Andrej Žemva, Andrej Trost, Carlos Miguel Correeia Da Costa, Armando Sousa, Germano Veiga, Héber Miguel Plácido Sobreira, Dragan Perakovic, Marko Periša, Ante Bilic Prcic

Published by InTech

Janeza Trdine 9, 51000 Rijeka, Croatia

Publishing Process Manager Iva Lipovic
Technical Editor InTech DTP team
Cover InTech Design team

First published October, 2015
Printed in Croatia

Additional hard copies can be obtained from orders@intechopen.com

Cutting Edge Research in Technologies, Edited by Constantin Volosencu
p. cm.
ISBN 978-953-51-2186-2

For EU product safety concerns:
IN TECH d.o.o., Prolaz Marije Krucifikse Kozulić 3, 51000 Rijeka, Croatia, info@intechopen.com or visit our website at intechopen.com.

Meet the editor

Constantin Volosencu is a professor in the Department of Automation and Applied Informatics at Politehnica University of Timişoara, Romania. He has performed research in the field of linear control systems, fuzzy control, neural control, control of electrical drives, sensor networks, distributed parameter systems, and fault detection and diagnosis. He is the author of several books and over 150 papers, manager of research grants, plenary speaker and committee member at international conferences, and a member of editorial boards for scientific journals. He was a research engineer at "Electrotimis" Enterprise in Timişoara, Romania, specializing in electrical drives and power ultrasonic equipments. He is also the author of 27 patents, with homologation of 30 prototypes and series zero.

Contents

Preface

This book responds to the great interest for innovation in the large domain of technologies. Manufacturing, electronic components, computers, development of the Internet, and healthcare are in need of advanced technologies that have been presented by researchers in this book. The book *Cutting Edge Research in Technologies* presents contributions by researchers with high expertise in the field, serving as a valuable reference for scientists, researchers, graduate students, and professionals. The subjects were chosen so as to cover the greatest advancements in the field, which are today at the forefront of technology. The authors were highly motivated to share the results of their new research for the benefit of the readers.

The book has five chapters covering the following subjects: information and communication technologies and services with the aim of improving the quality of life and the mobility of users; localization technologies for deployment of mobile robots in dynamic environments; embedded video processing circuit design flow in the Python language; data communications and networking; and textile weaving.

In the domain of information and communication technologies and services with the aim of improving the quality of life and the mobility of users, some possibilities of applying ICT to improve safe movement of blind and visually impaired persons are presented. In order to achieve greater information and safe movement of the user in the environment, identification and definition of the relevant parameters necessary to define the user's requirements are made, as the basic precondition for the design of new information and communication services.

In the field of localization technologies for deployment of mobile robots in dynamic environments, a 3 DoF/6 DoF localization system for low computing power mobile robot platforms is presented. The evaluation of the developed dynamic robot localization (DRL) system in three computing platforms is performed. The self-localization system is also able to perform mapping of the environment with probabilistic integration or removal of geometry and can use surface reconstruction to minimize the impact of sensor noise.

In the domain of embedded video processing circuit design flow in the Python language, an overview of video processing requirements, programmable devices used for embedded video processing, and the components of a video processing chain is presented. A novel design flow for generating customizable intellectual property cores used in streaming video processing applications is proposed.

In the domain of data communications and networking, an analysis of data link control protocols is presented, in the presence of error.Finally, in the domain of textile weaving, an overview of the historical facts of the art of hand weaving with patterned fabrics with multiple effects is provided. In addition, a way to determine the technique of weaving and construction parameters of each fabric pattern is provided.

I hope that specialists in the field read more, as the aim of this book is to educate them and get them interested in pursuing in-depth study.

I am happy to see this book being published, as it is source of great learning in which researchers have shared their valuable experiences. It also covers the state of the art in a world that needs more technological advancements to prosper.

I would like to thank all the researchers for accepting the invitation to contribute to this book and hope that it will leave its mark in the field.

Prof. Constantin Volosencu
'Politehnica' University of Timişoara
Romania

Chapter 1

Possibilities of Applying ICT to Improve Safe Movement of Blind and Visually Impaired Persons

Dragan Peraković, Marko Periša and Ante Bilić Prcić

Additional information is available at the end of the chapter

http://dx.doi.org/10.5772/61080

Abstract

Today's level of the development of information and communication technologies enables the implementation of assistive technologies that can contribute to improved mobility of the persons with impaired vision (users that move along the traffic network). The user in this research has the role of a pedestrian moving along the traffic network, using information and communication technology (ICT) solutions and services for the purpose of information about the surrounding and navigation. In order to achieve greater information and safe movement of the user in the environment, one has to identify and define the relevant parameters necessary to define the user's requirements, as the basic precondition for the design of new information and communication services. The analysis of the most used application solutions for mobile terminal devices showed the failure in providing precise information to the user, designing of functionality, structure of information and education of the users about the new solutions and services. The downsides of the current applications have served as the basis in defining the recommendations for the development of future applications, with the aim of increasing the user safety. Proper structure of information allows the user a faster and easier search of relevant information and information methods while moving along the traffic network elements. Therefore, the recommendations in designing future solutions and services based on possible technologies of short coverage area (RFID, NFC, Bluetooth, WiFi, RTLS) have been defined. These technologies allow communication connectivity of the users, other traffic entities and the entire traffic surrounding into a unique whole by using the principle of Internet of Things (IoT).

Keywords: Internet of Things, Assistive technology, Cloud computing, Mobility, Navigation

1. Introduction

According to the data of the World Health Organisation, there are today 285 million people in the world with impaired vision, out of which 39 million are regarded as blind while 246 million are regarded as partially sighted [1]. According to the latest report about the disabled persons in the Republic of Croatia, there are 17,428 persons with impaired vision, which is 3.4% of the total number of persons with disabilities [2]. According to the mentioned literature, there are 1,961 persons with impaired vision in the City of Zagreb. Out of this number, 185 blind persons (users) move every day using the white cane aid. This paper presents the analysis of implementing the information and communication (IC) technologies and services with the aim of improving the quality of living and the mobility of the users. The user is in the role of a pedestrian moving along the traffic network and their aim is to get precise information about their location and environment. From the definition of the pedestrian and categorisation, the disabled person represents the more endangered group in the traffic system, which includes also persons of impaired vision [3]. Independent moving of the users is based today exclusively on the use of aids such as: the white cane and guide dog. Infrastructure such as the traffic network element is also an important parameter in the function of user orientation and navigation. The accessibility elements are part of the infrastructure which can also be implemented into the intersections and public urban transport stops (trams, buses, taxi stands) [4]. Today's time of new technologies and services may provide the users better information, as well as adjustment of the traffic environment to the users' requirements. The implementation of information and communication technology according to the user's requirements allows the user to overcome social and infrastructural barriers which is the basic aim and purpose of the assistive technology [5, 6]. Some of the implementations of the assistive technologies are reflected in the use of IC solutions for navigation and guiding of the users along the traffic network. The mentioned solutions are based on the global navigation satellite systems (GNSS), data transfer systems in mobile networks (GPRS, EDGE, UMTS, etc.) and geographic information systems (GIS). The user interface is an important component in such solutions because it has to be adapted and completely accessible through its functionalities. The analysis of availability and characteristics of the GPS systems reflects the basic problem in such solutions, the error in the information provided to the user about the location [7–9]. The drawbacks can be compensated by the implementation of other technologies (RFID, WIFI, NFC, Bluetooth, RTLS) for the sake of locating and navigating the user [10,–12]. Recognising the elements of traffic intersections as part of the traffic network can be done by implementing the points of interest (POI) within the mobile applications. POI marks have the role of informing the user about a facility which is located in their vicinity [13]. The information about the location and the environment of the user represents the basic parameters in defining the user requirement for the route planning. By collecting all the relevant data into a single information system can be done by implementing the conceptual model based on Cloud Computing (CC) platform [7]. CC is used as a platform in services for the recognition of traffic intersections, as well as in recognising the pedestrian crossings and informing of all stakeholders [14, 15]. The described research results in this paper will form the basis for the definition of recommendations and guidelines for the introduction and implementation of new user-tailored IC technologies and services.

2. Models of assistive technology systems

The development in technology in the past two decades has opened new possibilities for the persons with impaired vision who successfully compete with visually healthy persons in all life segments [5]. This competition could not be achieved without using assistive technologies. One of the most acceptable definitions of assistive technologies was provided by the Faculty of Medical Sciences, University of Campinas, Campinas, Brazil, which says that assistive technologies are an interdisciplinary field of knowledge which encompasses products, means, methodology, strategies, skills and services whose aim is the development of the functionality of the persons with impaired vision regarding autonomy, independence, quality of living and social inclusion. Similarly, the US Technology Related Assistance for Individuals with disabilities Act (1988) defines assistive technologies as any item, part of equipment, i.e. system whether acquired commercially, modified or adapted, that is used for upgrading, maintenance or improvement of functional abilities of the persons with disabilities.

The role of assistive technologies in levelling the possibilities of accessing information to handicapped pupils lies primarily in reducing the effect of sensory impairments, and within the frame of the social model the aim of assistive technology is to overcome the gap between the things that the persons with disabilities want to do and the things that the current social infrastructure allows them to do [19]. Assistive technologies consist of the equipment, devices and systems that can be used to overcome social, infrastructural and other barriers encountered by the persons with disabilities and those that prevent them from having equal participation in all aspects of the society [5].

The widely accepted overview of the assistive technology is represented by ISO-classification of technical aids (ISO 9999 Technical aids for the disabled), which has been adopted by CEN (European Committee on Normalisation) for information exchange about technical aids. This classification includes:

- aids during treatment and therapy;
- orthoses and prostheses;
- aids for mobility which enable personal mobility or transport;
- home aids and equipment;
- equipment intended for adjustments in the house;
- aids for communication, information and signalisation;
- aids for handling other items;
- aids and equipment for environmental improvement; and
- recreation aids.

Access to information is becoming all the more important in the life of any person, and it is especially important in case of persons with impaired vision. The majority of information is obtained through visual and auditive channels, and if information is not available in alternative formats and/or technology does not allow access to the information, the persons with sensor impairments experience reduced access to information.

In spite of all technological advancement, accessing information remains an almost unreachable aim for visually impaired persons, limiting thus their opportunities for employment, education, leisure and independence [16] [17]. Constant barriers prevent independent, reliable and timely access to information. The same authors state that the majority of sources of information (such as, e.g., newspapers, magazines, and TV programs, PCs) rely on visual channels and visual accessibility such as black print or application of video displays. In their opinion, the design and implementation of assistive technologies would provide equal access to information for the visually impaired persons.

The technology is one of the strongest allies to the visually impaired persons in overcoming the negative effects of visual impairment [18]. The technology is related to successful education and positive change in attitudes and has potential in reducing the influence of certain negative consequences brought by visual impairment. Closely connected with education, assistive technology makes it possible for the visually impaired persons to become more successful, function as equal members of the society and develop their personal self-respect.

The main strength of assistive technologies is balancing of the possibilities of accessing information as well as other social possibilities on equal level like the visually healthy persons. The way they will manage the information and use them depends on the users of assistive technologies themselves.

The users of assistive technologies also differ regarding their characteristics, interests, skills, values and level of impairment. As a result of great diversity in the requirements of end users, applications and context, there is need for a simple, efficient and unique framework.

The aims of the framework are as follows:

- implementation in any assistive technology system;
- classification of the assistive technologies system;
- defining the basic structure of assistive technologies system and its implementation in further analyses in device specification;
- development of new assistive technologies systems in order to satisfy the end users;
- support to the process of providing assistive technologies to a certain user with the aim of the user accepting the solution, and
- possibility of understanding the method of functioning of the system for researchers and engineers in social context.

Due to a very complex situation in the real environment in which the end user is often present, the access during modelling of the assistive technology system model often cannot satisfy all the aims of the framework; therefore, the focus is on two main approaches. The first is satisfying the requirements of the end user with adequate assistive technology and measuring the outcomes of using assistive technology and the second is development of general framework for device analysis.

The satisfying of user requirements of the end user with adequate technology is related to the methodologies of the quality of living. Today's studies place greatest emphasis on the development of technologies whereas very little is written about the very evaluation of their effects. Measuring the outcomes of applying assistive technologies requires consideration of technology and the users who use it which means that the traffic and technological solution for the navigation and guidance of the blind and visually impaired persons need to be placed in the context in which it is used. The assessments of the effects of assistive technologies can be divided into two categories: objective assessment (measuring characteristics, how fast the user can overcome a barrier or find data using the screen reader, etc.) and subjective assessment (survey of the user satisfaction with the proposed technology).

Methods for the assessment of the quality of living of the blind and visually impaired persons:

- method of assessing adequate technology and user (matching person and technology – MPT);
- method of assessing individual efficiency of assistive technology (individually prioritised problem assessment – IPPA).

The mentioned methods allow the defining of the currently available information and communication technology and services, whose purpose is to determine the location of the user, as well as precise guiding and navigation of the user to the destination.

Method of assessing the adequate technology and the user is a procedure used to determine the outcomes of implementing adequate technology to the user in a defined environment. It consists of three main components: user, technology and environment.

The assessment procedure is carried out together with the user through six steps out of which three steps refer to the questionnaire, and other three to the discussion about the outcomes and activities that need to be undertaken.

The method of assessing individual efficiency of assistive technology or IPPA assesses the efficiency of implementing assistive technology by determining the level to which the problems and barriers the user encountered in everyday activities have been reduced. The assessment is based on identification of the seven barriers by the user in performing everyday activities, which could be reduced by using assistive technology.

To design the assistive technology system, it is necessary to know the basic models such as: HAAT (human activity assistive technology) model and CAT (comprehensive assistive technology) model [5].

2.1. Human Activity Assistive Technology (HAAT) model

This model, according to Cook and Hussey (2002), is an example of development of the general structure which is used for the analysis, synthesis and development, but it excludes connection of the device and the user. The definition of the assistive technology system allows the user to perform activities in the context of social environment with possible assistance of some of the assistive technologies.

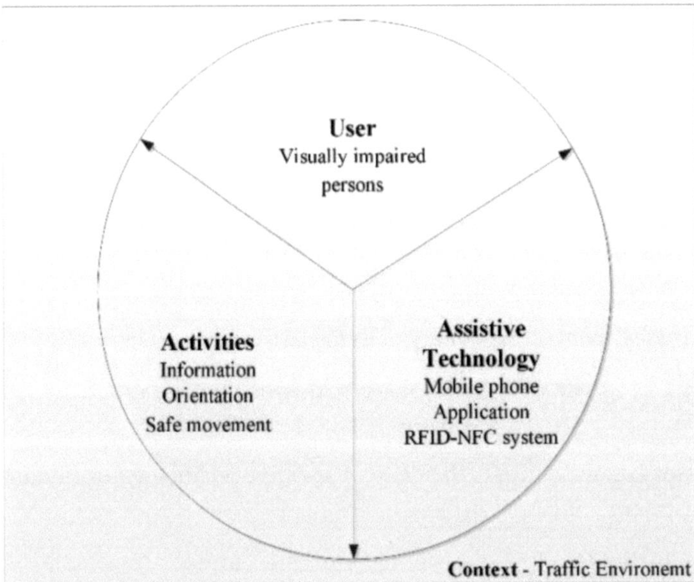

Figure 1. HAAT model according to Cook and Hussey

Figure 1 presents all the components of the model with the user who is in the centre of the system. The HAAT model system consists of four components:

- Context – social frame and physical surrounding in which the person and the assistive technology are functioning;
- Person – user in the centre of the model, which has the characteristics of a sensor input, and the power of central processing and motoric output;
- Activities – procedure, work or task performed by the user (the entire model depends on them), and
- Assistive technology – external assistance used to overcome contextual barriers and hindrances.

2.2. Comprehensive Assistive Technology (CAT) model

The model of final (more detailed) assistive technology has come out of the HAAT model. The structure of the model is in the shape of a tree, with limited number of variables on each branch. This display makes the model extremely understandable (Figure 2).

The highest level consists of four components that define the system of assistive technology:

- User – centre of the system;

http://dx.doi.org/10.5772/61080

- Context – environment in which assistive technology is used;
- Activities – which are to be used, and
- Technology that is used.

The aim of this model is to identify the drawbacks of implementing assistive technology. The result of this is the development of a system in the areas in which there is currently no elaborated approach or increase in the capacity of the existing system so that the user would have more options.

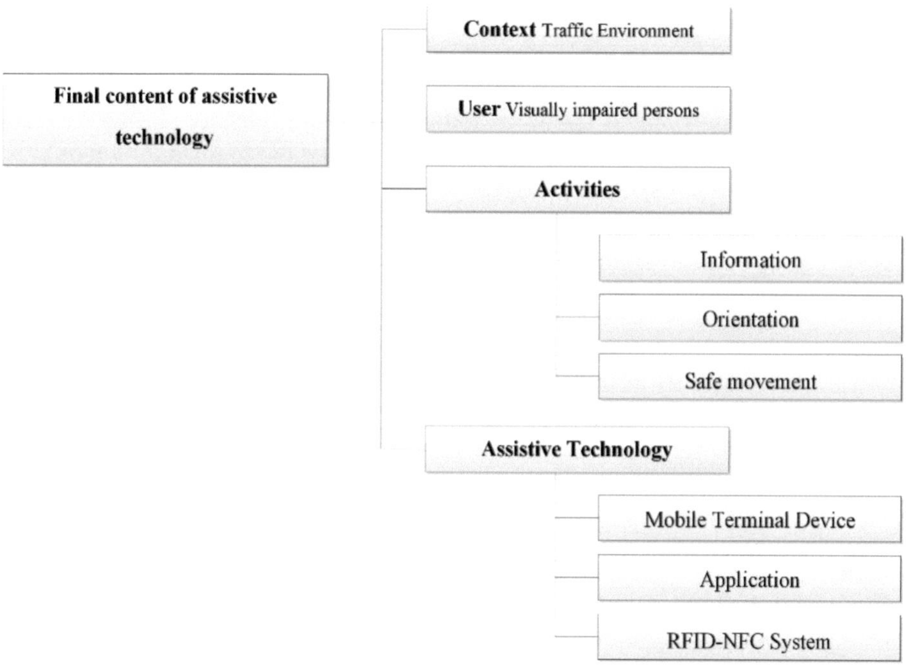

Figure 2. Presentation of CAT model

As an example one can mention the failure of installing audio signals at all signalised intersections, but rather only on some of them, making thus this system an assistive technology system. If all signal-controlled intersections were equipped with audio signalling devices, this would then represent a standardised solution.

3. Study of implementation of assistive technologies for the purpose of informing and navigating the users

Movement of the blind or visually impaired person is almost impossible today without the aids (white cane), which is not just an aid in moving, but also a symbol of the blind or visually impaired person. The idea for the usage of a white cane as the aid for moving was first invented in 1921 by James Biggs from Bristol (Great Britain). A decade later, the white cane was recognised in the society as an aid for the blind. In the USA, the implementation of this aid started in 1930 (Lion's Clubs International) with a black cane. After a lot of criticism because the persons using it were not sufficiently conspicuous, a white coloured cane started to be used. Since then, the white cane has become the most used aid in the function of orientation and movement of the blind and as such it has become the main element of assistive technology. The performed analyses of development of the current technology and devices have defined the relevant parameters for independent moving and information about the environment of the user moving along the traffic network. The presentation of the analysis of availability of mobile terminal devices and applications included participation of the users of the association HUPRT (The Croatian Society for Promotion and Development of Tiphlotechnology[1]). The procedure included participation of 16 users from the Society who use mobile navigation applications in their movements along the traffic network of the City of Zagreb. Testing of applications and mobile terminal devices (MTD) was carried out in the duration of 14 days at the traffic intersections in the City of Zagreb.

The presented research results included also, apart from the users of applications for guiding and navigation, the participation of other users from the HUPRT Society, with the aim of assessing the availability and adjustment of MTDs.

3.1. Analysis of availability of mobile terminal devices

In the analysis of MTD availability, the users have assessed the following: currently used operating systems, type of GPS receiver, input–output units, possibility of voice management of applications and the device itself. Table 1 shows the analysed characteristics of MTDs; the selected devices represent the most used group of currently used MTDs.

According to the analysed data, the most important parameter for the users in the hardware part of the equipment is the existence of a keyboard as input unit. The output units (voice) mostly used are TTS applications such as: Mobile Speak, TalkBack, Talks and those integrated in the operating system. The operating systems that have been analysed are important from the aspect of accessibility of applications used by the users.

The applications are analysed according to parameters that are listed in Table 2. The users have evaluated the importance (1 – not important, 5 – very important) of individual functionalities.

Loadstone GPS – application is free, i.e. it represents open-source variant GPS navigation application, which has been specially made for the blind or visually impaired users. The application operates with Symbian platform of series 60 and it can be connected with different GPS modules, either external ones or those installed in the mobile device. Loadstone does not use ready-made maps for movement and navigation, but rather the users themselves need to define the maps and movement routes that, after having been defined, can be sent to Loadstone central server so that other users may also use the made maps or routes. The application can

1 Joint term for all aids used by the blind and visually impaired persons

DEVICE MODEL	OPERATING SYSTEM	APPLICATION	GPS RECEIVER	INPUT UNITS	OUTPUT UNITS
Nokia E51	Symbian OS 9.2	Loadstone	A-GPS external – Iblue 737A	Keyboard	Talks
Nokia 6220 Classic	Symbian OS 9.2	Loadstone Nokia maps	BT-Q818XT 66-channel	Keyboard	Talks
HTC Mozart	Microsoft Windows Phone 7	Outdoor navigation	Integrated, A-GPS	Touch screen	Voice navigation
Sony Ericsson Xperia mini pro	Android OS, v4	Intersection Navigation WalkyTalky	BT-Q818XT 66-channel	Touch screen	TalkBack
Nokia C7	Nokia Belle OS	Nokia maps	Integrated, A-GPS	Touch screen	Mobile speak
HTC Vario 4	Windows Mobile 6.5	MobileGeo	Prestigio Bluetooth GPS	Combined	Mobile speak

Source: [20]

Table 1. Presentation of analysed characteristics of mobile terminal device components

operate off- or online for which the connection to Internet is necessary, so that the decision of the usage lies on the end user. The advantage of this application is the language support (Croatian) and functioning with Symbian screen readers that include Talks and Mobile Speak.

Outdoor Navigation – Windows phone 7 application has the possibility of selecting the maps (Google maps, OpenStreetMaps or OpenCycleMaps) for usage. The application also supports offline/online operating mode, which is extremely important for the users (social aspect). A large number of possibilities such as the possibility of independent input of points of interest (import KML[2] and Geocaching of LOC files), defining of SOS calls that may be in the form of SMS message or e-mail contact, sharing of defined routes of movement by means of online Facebook account or by sending e-mails. The application is additionally equipped by trip computer which enables measuring of the average movement speed, distance travelled, measuring of the altitude and integrated digital compass. The downside of the application is the impossibility of using any screen readers and the price.

Mobile Geo – application can be installed on any mobile device supported under Windows Mobile platform. Mobile Geo cooperates directly with the screen reader Mobile Speak for smart phones and enables users to use mobile phones with installed GPS modules or to connect them with other commercial modules. Using GPS solutions developed in the Sendero group, enables

2 Keyhole Markup Language – files read by Google Earth application

	APPLICATIONS					
	LOADSTONE GPS	OUTDOOR NAVIGATION	MOBILE GEO	INTERSECTION EXPLORER	NOKIA MAPS	WALKYTALKY
Language – HR	Yes	No	No	No	Yes	No
Operating mode (*offline/online*)	Yes	Yes	Yes	No	Yes	No
Type of map	Google maps	Bing maps OpenStreetMaps OpenCycleMaps	TomTom	Google maps	Nokia maps	Google maps
Automatic recognition of usage (pedestrian/vehicle)	No	No	No	No	No	No
Voice management	No	No	No	No	Yes	No
Input of points of interest (POI)	Yes	Yes	Yes	No	Yes	No
Starting mode	Slow due to external GPS	Fast	Slow	Slow due to external GPS	Fast	Fast
***Multitasking*[1]**	Yes	Yes	Yes	Yes	Yes	Yes
Method of creating the route	External (using computer)	Yes	Yes	No	No	No
Possibility of automatic creation of return route	No	No	No	No	No	No

Source: [20]

Table 2. Presentation of analysed functionalities of navigation applications and their accessibility

the Mobile Geo user great portability and flexibility in providing diverse information, using the installed maps in the memory of the mobile terminal device and at the same time enabling the formation and 100% control of new routes or upgrade of the existing ones. The user has the possibility for central transfer of licences which means that, if the user wants to change the mobile terminal device, the licence can be transferred to the new one without additional costs. The downside of the application is the unavailability of the digital map of the Republic of Croatia, and Mobile Speak for smart phones which operates on Windows mobile platform also has no voice support for the Croatian language.

Intersection Explorer – application which is exclusively intended for the blind and visually impaired users, does not have the function of user navigation, but rather provides the user with information about the location of the traffic intersection. It operates on the principle of Google tools Street view which provides the users virtual research of the locations and orientation by means of panoramic images recorded at the street level. A blind or visually impaired person with this study has the possibility of easier perception of the environment (traffic intersection) which surrounds them. The application operates exclusively on mobile terminal devices that use the operating system Android.

WalkyTalky – application which is also intended for the blind and visually impaired persons, operates together with the application Intersection Explorer (Android). The intention of the application is to guide and navigate the users to the final destination by using Google maps. The application allows the search of points of interest, but has no possibility of their input. The advantages of the application are voice support integrated in the application.

Nokia maps – application which is mainly used by the newer generations of Nokia mobile devices (Nokia Belle OS and Nokia Anna OS), but its maps can be loaded today also for iOS and Android operating system. Nokia maps allows storing and sharing of routes used by the user via social networks or by sending electronic mail. A detailed overview of the points of interest and their input is defined according to the user's requirements. Information about public transport is a possibility not provided by any of the analysed applications, and they are of great assistance to the blind and visually impaired persons. The application can be downloaded to the mobile terminal device by creating a user profile which is also a great advantage in relation to the applications that are charged.

The analysis of applications has identified the drawbacks such as the impossibility of automatic creation of return route and language voice navigation in Croatian if the user has no installed voice application. The method of starting and configuration in some applications is very complicated which makes it difficult for the blind or visually impaired person to use the application. Automatic recognition of the mode of usage, e.g. if the user goes on foot along one section and after that enters the public transport vehicle, is not enabled by any of the analysed applications.

3.2. Analysis of current technologies in the function of locating and informing of users

In the closed spaces and premises which disturb the GPS signal, i.e. where the possibility of determining the position of the user by implementing the GPS system is difficult or almost impossible, the positioning may be performed by the application of the mentioned technologies:

- RfID;
- Bluetooth;
- NFC - (Near field communication);
- Wireless LAN;
- RTLS; and
- **Locating by means of base stations:**
  - GSM – Global System for Mobile Communications;
  - UMTS – Universal Mobile Telecommunications System, and
  - LTE – Long Term Evolution.

The main characteristics in user positioning of the mentioned technologies are presented in Figure 3. The technologies are analysed with the aim of obtaining precise position of the user using it. The basic data about RfID technology is maximal working distance, which affects the advantage of this technology [22]. Technology of connecting Bluetooth and NFC have their advantages and are recommended to obtain information up to a maximum of 0.20 [m], and their advantage is low energy consumption. Wireless technology is also reflected in the advantages of the transfer speed and the security of data transfer [23, 24]. Locating by means of base stations has no significant advantages in this group of users, and the disadvantage is insufficient precision in determining the location.

Important characteristic of individual technologies is the capacity of data stored in tags (RFID, NFC), which is important from the aspect of user information. In case of RFID technologies, the mentioned data depend on the performance mode, therefore in case of Passive tag the data capacity is from 48 to 736 bytes. Active tag has the capacity of 64 to 32 KB, and Read-only which has the capacity of 20 bits.

Bluetooth technology depends on the version on the mobile terminal device, and it is exclusively used for information transfer to the user. The transfer speed depends on the versions, the latest version Bluetooth SIG (V 4.0LE) allows speed of 25 Mbps. NFC technology allows data transfer speed of up to 424 kbit/s, parallel to RFID technology represents Point-to-Point communication, and a scope smaller than 0.2 m. Wireless technology provides transfer speeds depending on the protocols; standard protocols operate at frequencies of 2.4 GHz (802.11b and 802.11g) and 5 GHz (802.11a) and allow transfer speeds of up to 54 Mbit/s.

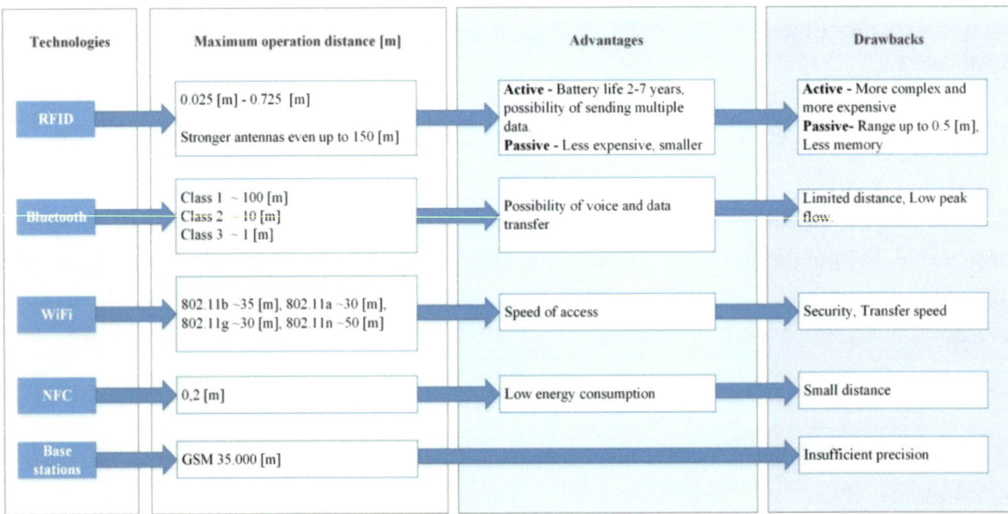

Figure 3. Characteristics of other technologies in positioning of the user [21]

Mobile terminal devices that have been adapted to the visually impaired persons are equipped with analysed technologies. The basic functionalities are used independently by 84% of interviewed users, whereas 16% of users have the problem due to lack of adaptation. The mentioned data are important since the most used mobile terminal devices equipped with the mentioned technologies (Bluetooth, Wireless, NFC, GPS) have been analysed.

For more precision in obtaining information about the user location who moves along the traffic network, it is recommended to use RFID technology. The mentioned technology is used to identify and inform the user about the state and environment of the traffic intersection. The user receives information by using the mobile terminal device.

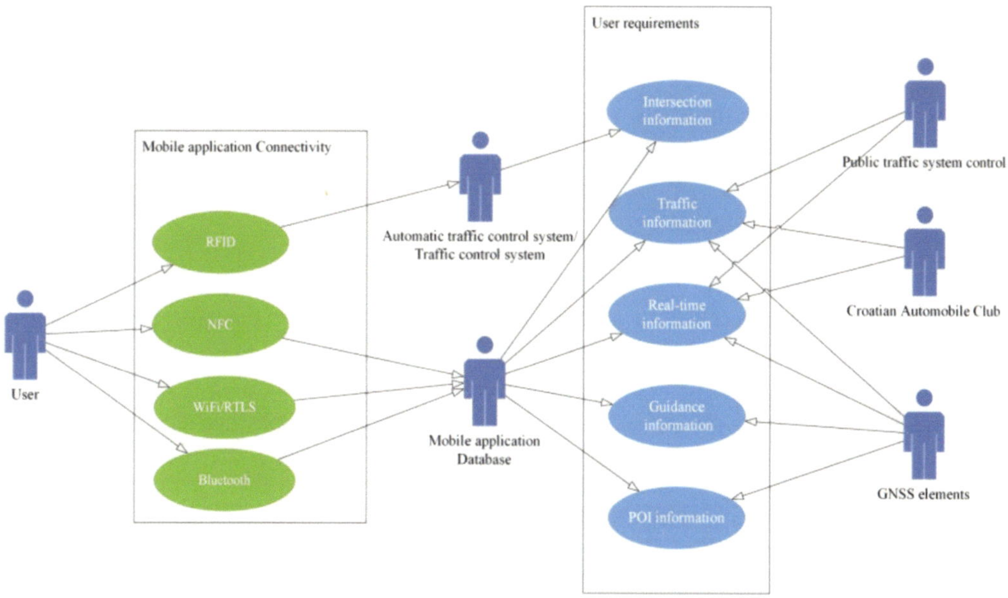

Figure 4. Preview of fulfilling of user requirements based on different mobile application connectivity [21]

The possibilities of applying individual technologies are presented in Figure 4, which presents in detail the connectivity between the user (user law) and all the stakeholders.

4. Defining of user request in the function of realising the aim of the assistive technology

The movement of users along the traffic network depends on a number of key parameters. The solutions that are today used in the function of guiding and navigation of users and information (according to user requests) have not sufficient presence of relevant parameters. The role of key parameters is to enable the adaptation of the model to users' requests providing the user with safe movement.

4.1. Identification of relevant parameters of guiding and navigating

The systemic approach enables identification of the key parameters which result from two scientific areas: field of technology of traffic and transport and the field of education and rehabilitation science.

For model optimisation, i.e. adjustment of individual elements of the traffic system, it is necessary to optimise function *Pz*, i.e. to evaluate single function variables:

Ku – quality of service;

Sr – contribution to solution standardisation;

Zp – implementation of law and regulations, and

E – education of users about new solutions.

From the mentioned conditions, the function of satisfaction is formed:

$$Pz = f(Ku, Sr, Zp, E) \tag{1}$$

By studying the needs of the users moving along the traffic network, the traffic parameters have been defined (field technology of traffic and transport), whereas the parameters from the scientific area of social sciences (field of education and rehabilitation science) were studied by the implementation of training of education and movement. The training of orientation and movement satisfy the following elements: moving across the open space, movement on internal polygon, and moving along the traffic intersection.

The results of the mentioned training are the definitions of relevant parameters which are the basis of further research in the area of designing models of guiding, navigation and information of users. Table 3 shows certain parameters depending on the scientific field.

Field of traffic and transport technology	Field of educational and rehabilitation science
Speed	Perception
Safety of movement	Orientation
Precise information	Independence
Time	Mobility
Landmark	Education of users

Source: [25]

Table 3. Parameters of guidance and navigation of the blind and visually impaired persons within the traffic network

Definitions of parameter from the presented table:

- speed – notion that defines the speed of user movement along the desired route (depending on the route and time);
- time – notion which describes the duration of the user movement along the desired route;
- safety of movement – undisturbed movement, so that the user acquires confidence in the proposed solution and gets the feeling of safety;

- precise information – information which raises the user's feeling of security, because if information is not precise, the movement of the user may be endangered;
- landmark – data which may give the user the information in which direction to continue moving, information about their location;
- perception – recognition of the environment in which the user is positioned due to which the information is memorised;
- orientation – managing in space (not equally expressed in all persons);
- independence – feeling that the proposed model allows the user moving with minimal assistance by another person;
- mobility – free movement of the user, i.e. ability of walking in a safe and coordinated way, and
- user education – education of the users about the method of using the system.

The conceptions that surround the user while moving along the traffic network are an important segment in the creation of the knowledge base. The relevant parameters of guidance and navigation of users while moving along the traffic network can be presented as the life-cycle of knowledge.

4.2. Defining user requests

By using the aids in the function of movement, the blind and visually impaired persons want to arrive from point A to point B in a safe, simple, efficient and independent manner. For that purpose, it is important to define users' requests that are based on relevant parameters. The users' requests can be divided into two categories:

- Basic information (use of current aids) and
- Expanded (additional information by using assistive technologies).

The basic users' requests consist of the following:

- Information on location;
- Information on guidance and navigation;
- Information on facilities surrounding the user;
- Information of audio character;
- Information about descending and ascending kerbstone;
- Information about the traffic intersection;
- Information on the method of traffic control (tactile line, traffic light system or something else);
- Information of the right moment to cross the street;

- Information on the system operation (breakdown of the system or upgrade); and
- Information of arrival to the destination.

The notion of identification of the user is defined within the zone of identification (their location) – depending on the size of the traffic intersection, the zone of user identification is defined. Informing the user about the location and navigation (shape of the traffic intersection and all its elements) includes an example when the user obtains from the system precise information about their location in the form of audio or voice information. The system navigates the user by tactile or voice information. Information of users about the facilities that surround them includes facilities that surround the user and which have to be in the identification zone. The facilities may be state institutions, banks, hospitals, cultural facilities of significance and other facilities that can be found in the user's surrounding. Enabling the actuation of the audio signalisation contributes to the raising of the level of the quality of living of citizens into the environment with the solution, mostly due to the noise produced by audio signalling devices at traffic intersections, the system allows actuation of audio information after user identification. At the moment of audio signal actuation, the user has to have enough time to orientate themselves towards their target. The management of the real-time information of the user is a service that informs the user with audio or voice information about the changes on their movement route. Example: if there are works on the pedestrian crossing and there is no possibility of crossing it, the user receives information about this, and receives suggestions about the alternative routes for safe movement. Provision of information according to a greater number of criteria and special points of interest is used if the user uses navigation application in their movement, the system provides information such as: selection of the shortest route, information input on the navigation map about the user environment, pre-announcement when arriving to the input point of interest (example: how many metres the user has to go to the defined point).

Provision of information about the direction of movement using tactile and voice information is possible by using the elements of accessibility during which the user receives information about their direction of movement. Voice information provides the user with the information about the size of the intersection, its elements (the number of lanes in which direction), existence of bus and tram stops. By applying the basic and secondary geographical parts of the world, the user gets information about navigation along their route. This is voice information and the user receives information in the following form: movement in EAST–WEST direction along the Harambašieva Street.

The logic structure of information allows the user a more suitable method of using the service; according to the most frequently used information, logic structure of the system allows easier approach to the most frequently used information. The most frequently used information is defined according to single user requests, which also provides the dynamics of the model itself.

Two-way information, data and voice communication with the user, provides the user with the possibility to define their route prior to starting their movement along the traffic network by using navigation applications, and during their movement, the user can independently enter the information that had not been entered before. For safety reasons, the information can be very important, for instance, if there are works in the direction in which the user is moving or some information at the intersection are incorrect.

Information of users, position precision of the user – by applying other technologies (RfID, Bluetooth, WiFi, RTLS or NFC) the user receives precise information about their position.

Automatic control of signal-controlled system, longer green phase for pedestrians – when identifying the user, the system allows longer duration of green phase for pedestrians. After the user leaves the system, the system returns to the state prior to user identification.

Another category of user requests represents a group that is based on the already described definitions, but supported by new ICT and services. Therefore, it is necessary to satisfy additional requirements:

- precise location of the user ~0.5 [m];
- precise information about the movement;
- user-friendliness of the mobile terminal device (keyboard and screen reader);
- possibility of creating priority information;
- economically affordable solution;
- possibility of creating pre-announcement prior to arriving to the destination;
- possibility of facility identification;
- possibility of identifying the traffic intersection, and
- selection of the device operation mode offline–online.

For everything mentioned, it is important that the system elements can operate in any weather conditions for the user safety. If the weather conditions allow some changes in the system operation, the user has to be informed.

5. Possibilities of applying advanced technologies in increasing the safety of user movements

The user requests described according to the possible operating modes can be integrated into one whole by applying the concept of Cloud Computing. For this purpose, the CCfB (Cloud Computing for the Blind) architecture has been defined and it makes it possible to combine all relevant information into one database [7]. An example of such architecture is presented in Figure 5 where the user is in the very centre of the system, and is surrounded by all relevant information necessary for safe and coordinated movement along the traffic network. The user access to relevant information is possible by implementing Web 2.0 technology, which allows adapted Internet and mobile application [26].

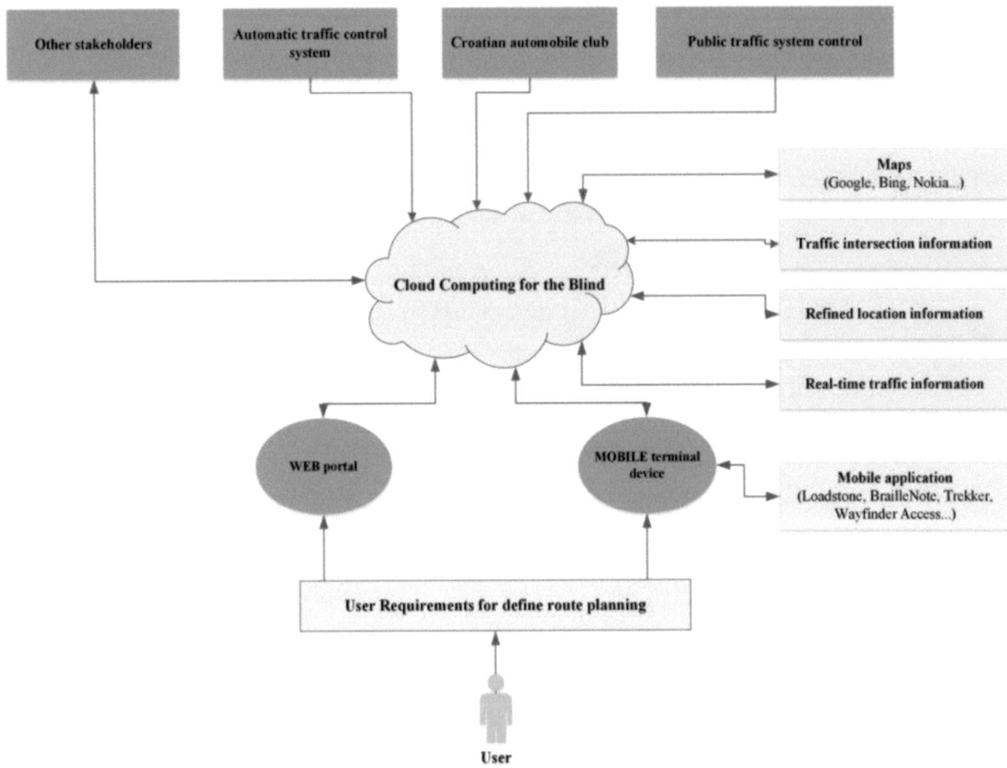

Figure 5. Simplified architecture of the Cloud Computing for the Blind system

Functionality of the presented model based on CC platform is shown in Figure 6. CCfB architecture is based on IaaS service model described by UML diagram of the Use case. IaaS model provides the possibility of storing data and computation abilities as standardised service via network. The data found in IaaS architecture are created and updated by the users, service provider and the third party (stakeholders). This allows delivery of services in SaaS or PaaS architecture in the public scenario, which has been defined depending on the user requirements.

The implementation model used in the presented solution is the Public Cloud, where the infrastructure in CCfB is accessible not only to the users but also other stakeholders (AuP system, HAK (Croatian automobile club), public urban transport system, traffic light control system). Public Cloud represents a model of open use by public, infrastructure an be managed, used and in the ownership of one or several business, public or state organisations (it exists at the service provider location). The mentioned approach, i.e. by applying CC platform in combining the described data, the dynamic scaling of the system depending on the users' needs and the requirements of the system itself is possible.

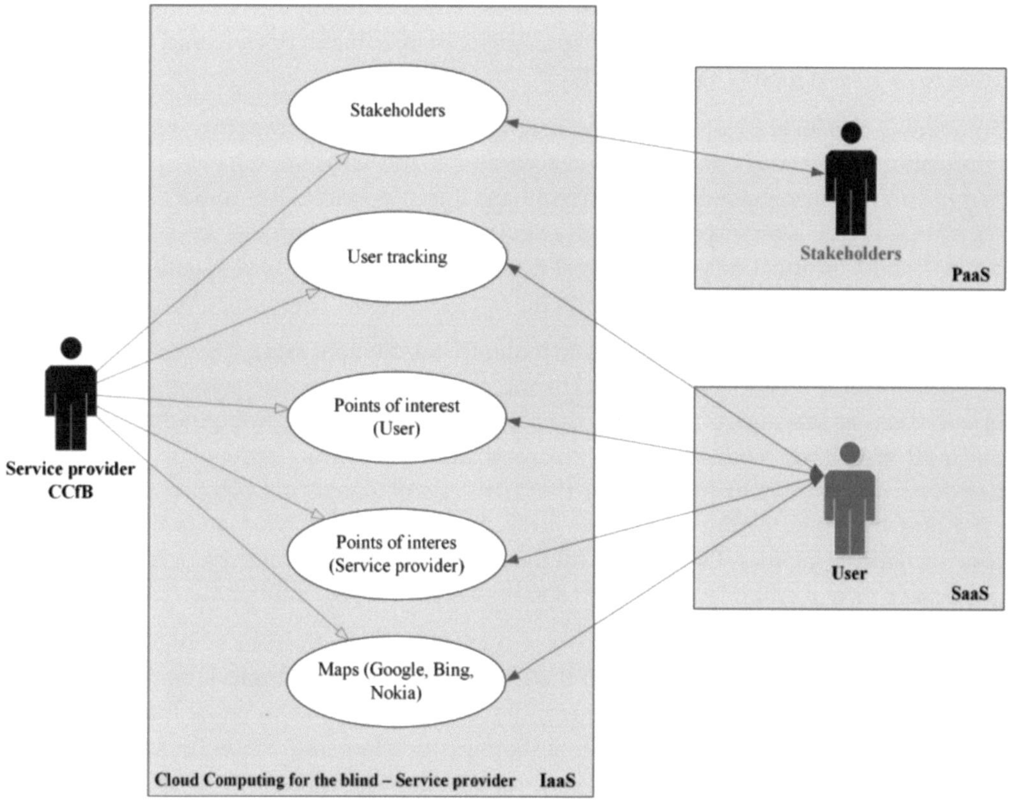

Figure 6. Functionality of conceptual model based on Cloud Computing platform [7]

The mentioned architecture elements (User tracking, Points of interest) allow creation of the user knowledge base, thus providing the system in future operation with independent decision-making processes.

The access to the contents of information, as already mentioned, is possible by using mobile application or by means of the Internet browser. Therefore, the MTD interface accessible to visually impaired persons is important to satisfy all the accessibility aspects "Design for Usability" [27]. According to the elements of universal design, the mobile application has to be available to all the groups of users, which means that it should not deviate in its design and possibilities from the standard solutions. The design has to be equally adapted for both the left-handed and right-handed persons. The usage flexibility is important when satisfying the user requirements, and for this purpose it is recommended during user installation to define the user profile so that the user would always have the requested information. Because of the possibility of connecting the application with the web interface, it is necessary for the mobile and web application to have satisfied standards in the selection of colours for the partially sighted persons, as well as the possibility of increasing or reducing the font size.

The contents of application must provide clear and understandable information, mostly because of compatibility for the screen readers. When defining the basic information, it is important to provide the users with the possibility of defining their own level of disability in order to be able to modify the information.

By defining the level of disability, the user can have better accessibility of the requested information, and it can be thus modified according to the user requirements. Image information has to be accompanied by description and the information has to be understandable, regardless of the user experience, knowledge, language knowledge or current level of concentration. Information provided by the application also has to have linguistic support for the majority of world languages.

The application design has to be such as to minimise the dangers, and to prevent consequences of incidental or unintentional activity. The information management elements need to be set in such a way as to reduce to minimum the danger and errors due to the application operation: the most frequently used elements; the most accessible ones; elimination of dangerous elements, isolated or covered. Insure the warnings of danger or possible error. Provide protection elements. Disable involuntary procedures in creating information that require full concentration of the user. Operation with the application should not present for the user any physical or mental effort, i.e. such effort should be reduced to a minimum.

The real-time passenger information, a service which informs the user by means of audio or voice information about the changes on their movement route. Example: if there are works on the pedestrian crossing and there is possibility of not being able to cross, the user receives information about this, and receives suggestions about alternative routes for safe movement. Two-way information, data and voice communication, with the user is important before moving along the traffic network. The users can create themselves their routes within the application, the information that is not entered can be entered by the users independently during their movement. The information can be very important for safety reasons; for instance, if there are works along the direction of the user movement or the information at the intersection is incorrect.

The accompanying contents that surround the user can be defined within the application as points of interest. The possibility of such information can be used if the user wants to receive information about the accompanying content such as cafés, restaurants, museums, shopping centres, hospitals, etc.

Because of the possibility of connecting with the web application, it is necessary to insure compatibility of the device. Current data about the number of used devices go in the favour of devices that have a keyboard, whereas there is less usage of the devices with touch screen as input units. Because of its input unit, iPhone device provides a keyboard as an additional component which is also a very good characteristic of the producer in approaching design.

Stakeholders presented in Figure 6 can be expanded with users who base their approach on volunteering. For this purpose, the goal is to enable complete functional service of providing assistance in situations when others need help, regardless of the level of degree and type of assistance to the end user [28]. Today's applicative solutions can be expanded for this purpose by modules that would connect the people who want to help and those who need that help. Today's development of the technology allows this without mediation of various organisations and societies and strengthens the feeling of unity and trust to public welfare.

The integration with geo-location services allows sending pre-defined SMS messages with the possibility of editing, i.e. automatic connection of pre-defined services and professional services and organisations. Using simple examples: crossing an intersection, passing along a rutted pedestrian path or entering a public transport means, it is obvious that such modules really help the people who want to provide assistance. Special focus is on the social component and further spontaneous development of the system towards a maximally wide range of services and forms of assistance.

For the purpose of informing the users, it is possible to use the short-range technologies (NFC and Bluetooth), as presented in Figure 7. Possibility of informing the user who is at a traffic intersection and uses mobile application on their MTD can be performed by using NFC or Bluetooth technology [29].

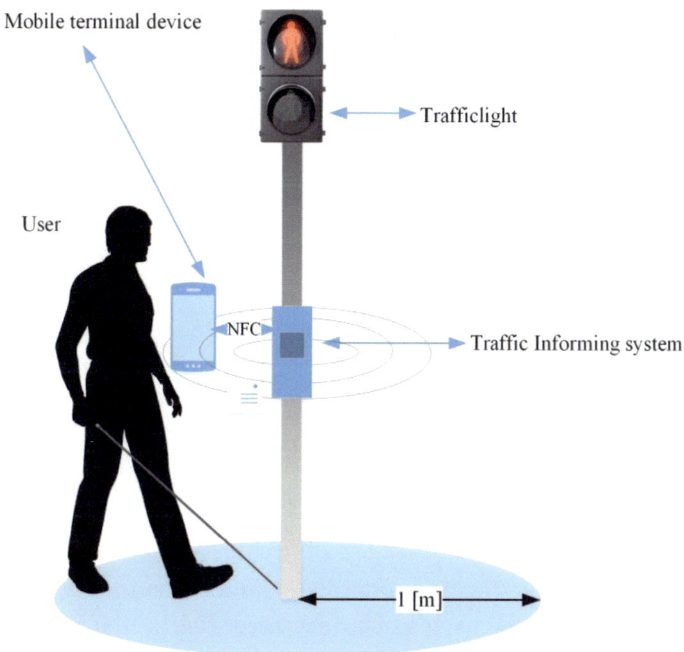

Figure 7. Possibility of implementing NFC technology for user information

Therefore, it is necessary to satisfy the criteria of the information structure in the very tag, as well as the type of tag. The information need to be simple and precise so as not to endanger the safety of user movement. The user requests also play a big role in creating information. The users' requirements create the priority information important for safe and independent movement of the users. Proper architecture of information provides the user with faster and easier search of relevant information, creation of movement routes, methods of information while moving along the traffic network elements and customised contents.

Information provided by the user information service contains all the elements of universal design, which is the basic principle for independent participation of the users in everyday life. This refers to the unbiased possibility of usage, equal methods of usage for all users, flexibility, conspicuity, low physical or mental effort and toleration of errors.

By using their mobile terminal devices, the users receive information in the following form:

Direction of movement north–south Šubićeva street, oblique pedestrian crossing, body posture 30° to the right, three lanes east–west, two lanes of tramway tracks, pedestrian island, three lanes west–east.

The mentioned information consists of the current location which is indicated by the cardinal points of compass (north, south, east, west and derivatives) and the name of the street which is within the user's environment. The information about the traffic intersection describes all the elements of which the traffic intersection consists with the description of the direction of the vehicle movement. The direction of the vehicle is also described by using the cardinal points of compass. If the intersection configuration has a certain trigonometric form or the pedestrian crossing is set at an angle, then these data also have to be included in the information provided to the user. The distance of the kerbstone is defined by the users using the aids (white cane or guide dog), where the mentioned information in NFC tag is of informative significance. According to statistical indicators, an increasing use of NFC technology is predicted and therefore in further development the mentioned solution is expected to find its application in real environment. The implementation of the mentioned technology is possible in several aspects, one of the examples being in the area of e-Health [30].

6. Conclusion

The mobility, accessibility, place of residence, transportation and education of the users are the basic conditions for independent and full participation of the disabled persons in everyday activities. The presented results of the performed studies in this paper say how much the new technologies can help in removing the barriers and hindrances in performing everyday activities of the disabled and senior persons and persons of poor mobility. The design and development of mobile applications for informing, guiding and navigating the users who move along the traffic network have to be done according to the recommendations and standards. The mentioned presentation of past research in this area can be used as the basis in defining the future guidelines about the methods and possibilities of providing new technologies. The new technology and its application according to the presented paper will certainly help in removing the barriers which include the target group of users.

As the basis of the decision-making processes, the new services use the user knowledge bases, with the aim of increasing the level of the users' quality of living. For this purpose, it is necessary to educate all the stakeholders such as the peripathologists, etc. about the possibilities provided by the new services.

Author details

Dragan Peraković[1*], Marko Periša[1] and Ante Bilić Prcić[2]

*Address all correspondence to: dragan.perakovic@fpz.hr

1 University of Zagreb, Faculty of Transport and Traffic Sciences, Zagreb, Croatia

2 University of Zagreb, Faculty of Education and Rehabilitation Sciences, Zagreb, Croatia

References

[1] World Health Organisation. Visual impairment and blindness [Internet]. [Updated: August 2014]. Available from: http://www.who.int/mediacentre/factsheets/fs282/en/

[2] Hrvatski Zavod za javno zdravstvo. Izvješće o osobama s invaliditetom u Republici Hrvatskoj [Internet]. 2014 [Updated: 2014]. Available from: hzjz.hr/wp-content/uploads/2013/11/Bilten_invalidi_2013.pdf

[3] Periša M, Vujić M, Sertić A, Anžek M. Pedestrian as a traffic entities in ITS environment. In: Anžek M, Hernavs B, Kavran Z, Petelin S. (eds.) ISEP 2009 – sustainable transport and mobility; 26.03.; Ljubljana. Ljubljana: Electrotechnical Association of Slovenia; 2009. p. R3, T.

[4] Periša M, Vujić M, Peraković D. Tactile network systems and intelligent applications for elderly and disabled travelers in public transportation systems. In: Anžek M, Hernavs B, Kavran Z, Meše P, Štern A. (eds.) ISEP 2010 – The role of ITS in the near future; 29.-30.03.; Ljubljana. Ljubljana: Electrotechnical Association of Slovenia; 2010. p. R19.

[5] Hersh M, Johnson M.A. (eds.) Assistive Technology for Visually Impaired and Blind People. 1st ed. London: Springer; 2008. p. 725.

[6] William CM. Smart Technology for the Aging, Disability and Independence. 1st ed. New Jersey: John Wiley & Sons; 2005. p. 395.

[7] Periša M, Peraković D, Šarić S. Conceptual model of providing traffic navigation services to visually impaired persons. Promet – Traffic & Transportation 2014;26(3): 209–18. DOI: 10.7307/ptt.v26i3.1492.

[8] Kammoun S, Macé J-MM, Oriola B, Jouffrais C. Designing a virtual environment framework for improving guidance for the visually impaired. In: Langdon P, Clarkson J, Robinson P, Lazar J, Heylighen A. (eds.) Designing Inclusive Systems: Designing Inclusion for Real-world Applications. 1st ed. London: Springer; 2012. pp. 217–226.

[9] May M, Casey K. Accessible global positioning systems. In: Manduchi R, Kurniawan S. (eds.) Assistive Technology for Blindness and Low Vision. 1st ed. Boca Raton: CRC Press, Taylor & Francis Group; 2013. pp. 81–105.

[10] Liao C, Choe P, Wu T, Tong Y, Dai C, Liu Y. RFID-based road guiding cane system for the visually impaired. In: Rau PLP. (eds.). Cross-Cultural Design. Methods, Practice, and Case Studies. 1st ed. Berlin Heidelberg: Springer; 2013. pp. 86–93. DOI: 10.1007/978-3-642-39143-9_10.

[11] Dudwadkar A, Gore A, Nachnan TI, Sabhnani H. Near field communication in mobile phones. Int J Eng Adv Technol. 2013;3(1):309–18.

[12] Burden M. Near field communications (NFC) in public transport. In: The Institution of Engineering and Technology Seminar on RFID and Electronic Vehicle Identification in Road Transport; 29.11.; Newcastle. Newcastle, UK: IET; 2006. pp. 21–38.

[13] Erdelj M, Razafindralambo T, Simplot-Ryl D. Covering points of interest with mobile sensors. IEEE Trans Parallel Distributed Syst. 2013;24(1):32–43. DOI: 10.1109/TPDS.2012.46.

[14] Angin P, Bhargava B, Helal S. A mobile-cloud collaborative traffic lights detector for blind navigation. In: Eleventh International Conference on Mobile Data Management (MDM); 23–26.05.; Kansas City. Kansas City: IEEE; 2010. pp. 396–401 DOI: 10.1109/MDM.2010.71.

[15] Yuriyama M, Kushida T. Sensor-cloud infrastructure – physical sensor management with virtualized sensors on cloud computing. In: 13th International Conference on Network-Based Information Systems (NBiS); 14–16; Takayama. Takayama: IEEE; 2010. pp. 1–8. DOI: 10.1109/NBiS.2010.32.

[16] Abner GH, Lahm EA. Implementation of assistive technology with students who are visually impaired: teachers' readiness. J Visual Impair Blindness. 2002;96(2):98–105.

[17] Augusto CR, Schroeder PW. Ensuring equal access to information for people who are blind or visually impaired. J Visual Impair Blindness. 1995;89(4):11–13.

[18] Edwards BJ, Lewis S. The use of technology in programs for students with visual impairments in Florida. J Visual Impair Blindness 1988;92(5):302–12.

[19] Johnstone C, Thurlow M, Altman J, Timmons J, Kato K. Assistive technology approaches for large-scale assessment: perceptions of teachers of students with visual impairments. Exceptionality 2009;17(2):66–75.

[20] Periša M. Dynamic Guiding and Routing of Disabled and Visually Impaired Persons in Traffic [dissertation]. Zagreb: University of Zagreb, Faculty of Transport and Traffic Sciences; 2013. p. 154.

[21] Periša M, Peraković D, Vaculik J. Adaptive technologies for the blind and visual impaired persons in the traffic network. Transport 2015;30(3). pp. 1-6. DOI: 10.3846/16484142.2014.1003405

[22] Finkenzeller K. RFID Handbook, Fundamentals and Applications in Contactless Smart Cards, Radio Frequency Identification and Near-Field Communication. 3rd ed. West Sussex: John Wiley and Sons, Ltd.; 2010. p. 480.

[23] Heydon R. Bluetooth Low Energy: The Developer's Handbook. 1st ed. New Jersey: Prentice Hall; 2013. p. 323.

[24] Stojmenović I. Handbook of Wireless Networks and Mobile Computing. 1st ed. New York: John Wiley & Sons, Inc.; 2002. p. 648.

[25] Periša M, Peraković D, Runjić T. Identification of relevant guidance parameter for Visually Impaired Persons in Traffic. In: Šakić Ž. (ed.) Korema 2011; 15–20.11; Zagreb. Zagreb: KoREMA, Croatian Society for Communication, Computing, Electronics, Measurement and Control; 2011. pp. 221–224.

[26] Periša M, Peraković D, Remenar V. Guidelines for developing e-learning system for visually impaired. In: Damm C, Penaz P. (eds.) Universal Learning Design, Vol. 2. ; 11.13.07.; Linz. Brno: Teiresias Centre of MU; 2012. pp. 167–173.

[27] Periša M, Jovović I, Peraković D. Recommendations for the development of information and communication services for increasing mobility of visually impaired persons. In: Penaz P, Červenka P, Hanouskova M, Masilko L, Pavliček R, Pecl J. (eds.) Universal Learning Design, Vol. 4; 09.12.07.; Paris. Brno: Masaryk University; 2014. pp. 103–112.

[28] Periša M, Jovović I, Forenbacher I. A conceptual applicative solution for helping people with reduced mobility. In: Mokrys M, Badura S. (eds.) RCITD 2014 Proceedings in Research Conference in Technical Disciplines; 17.21.11.; Zilina. Zilina: EDIS – Publishing Institution of the University of Zilina; 2014. pp. 82–85.

[29] Peraković D, Periša M, Jovović I. Near-field communication technology for informing blind and visually impaired persons when moving through traffic intersections. In: Vaculik J. (ed.) Zborník prednášok a príspevkov DoNT 2014; 23.10.; Žilina. Žilina: Žilinská univererzita v Žiline v EDIS; 2014. pp. 53–60.

[30] Dhar S, Dasgupta A. NFC technology: current and future trends in India. In: 2014 International Conference on Contemporary Computing and Informatics (IC3I); 27–29.11; Mysore. Mysore: IEEE; 2014. pp. 639–644. DOI: 10.1109/IC3I.2014.7019680.

Chapter 2

3 DoF/6 DoF Localization System for Low Computing Power Mobile Robot Platforms

Carlos M. Costa, Héber M. Sobreira, Armando J. Sousa and Germano Veiga

Additional information is available at the end of the chapter

http://dx.doi.org/10.5772/61258

Abstract

Mobile robot platforms have a wide range of hardware configurations in order to accomplish challenging tasks and require an efficient and accurate localization system to navigate in the environment. The objective of this work is the evaluation of the developed Dynamic Robot Localization (DRL) system in three computing platforms, with CPUs ranging from low to high end (Intel Atom, Core i5, and i7), in order to analyze the configurations that can be used to adjust the trade-offs between pose estimation accuracy and the associated computing resources required. The DRL is capable of performing pose tracking and global pose estimation in both 3 and 6 Degrees of Freedom (DoF) using point cloud data retrieved from LIDARs and RGB-D cameras and achieved translation errors of less than 30 mm and rotation errors of less than 5° when evaluated in three environments. The sensor data retrieved from three testing platforms was processed and the detailed profiling results were analyzed. Besides pose estimation, the self-localization system is also able to perform mapping of the environment with probabilistic integration or removal of geometry and can use surface reconstruction to minimize the impact of sensor noise. These abilities will allow the fast deployment of mobile robots in dynamic environments.

Keywords: Self-localization, point cloud registration, pose tracking, point cloud library, robot operating system

1. Introduction

Autonomous mobile robots capable of operating in dynamic environments have a multitude of applications and can be used to improve the overall speed and efficiency of a wide range of jobs. They can also cooperate with humans to accomplish complex tasks and are able to perform structured and repetitive movements with high precision.

Navigation within a dynamic environment requires a robust and accurate self-localization system in order to know where the robot is and which path it should follow to reach the location where it is expected to perform its tasks. The estimation of this global pose can use a varied number of techniques and technologies and may rely on proprioceptive knowledge that the robot has about itself (such as odometry), may incorporate exteroceptive information [1] retrieved from sensors (such as LIDARs, RGB-D cameras, sonars), or may even use infrastructures that are external to the robot (such as GPS). Nevertheless, for fast deployment of autonomous mobile robots on indoor environments, most localization systems rely on a combination of both proprioceptive and exteroceptive information to estimate the robot pose, given that external infrastructures are expensive for large operation areas and don't have either the coverage or the precision required for robot docking operations.

The most used self-localization systems can be categorized as probabilistic pose estimation methods, point cloud registration methods, or feature registration methods. Kalman filters such as the Extended Kalman Filter and the Unscented Kalman Filter [2] along with the particle filters [3] are the most used probabilistic methods but they rely on sensor models, and as such, it is hard to achieve very accurate pose estimations when the models are not well defined or when they change over time depending on the environment on which the robot is moving (which is the case of the odometry model). Point cloud registration methods that rely on the Iterative Closest Point [4, 5] or Normal Distributions Transform [6] can achieve very accurate pose estimations but require an initial alignment of the sensor data with the known map in order to successfully converge. Feature registration algorithms such as the ones presented in [7] and [8] don't require an initial alignment, but need distinctive geometry in the environment in order to successfully estimate the robot pose and are very computational intensive.

The DRL system combined feature registration methods with point cloud registration algorithms in order to provide high accuracy pose tracking along with global pose estimation when the robot starts its operation or becomes lost in the environment. This approach allowed the creation of a more efficient, accurate and robust localization system that uses the most appropriate types of algorithms depending on the knowledge that it has about the robot estimated pose in the environment. Each of these pose estimations have information about the registration of the sensor measurements with the reference point cloud, which includes the percentage of correctly registered sensor points along with their root mean square error and spatial angular distribution. The distribution of the initial poses are also given (when the localization system resets pose tracking, which occurs when the robot starts its operation without knowing where it is in the environment or when it becomes lost) in order to allow a navigation supervisor to detect if the estimated pose is ambiguous (when there are other known areas of the environment with similar geometry) and plot a path to disambiguate the estimated robot location. Besides pose estimation in 3 and 6 DoF, the DRL system can also incrementally build the map of the environment using probabilistic integration or removal of geometry along with surface reconstruction (to reduce the impact of sensor noise). This allows the robot to update an accurate representation of the environment while it performs its tasks, giving the possibility to explore unknown areas or plot paths through regions that where once unavailable.

The DRL system was evaluated using three testing platforms on three different environments and the sensor data was recorded into rosbags in order to allow the tests to be executed and profiled on three different CPUs (from low to high end). The test results showed that the DRL system was able to perform pose tracking with 5–30 mm of mean translation error and 0.4–5.0° of mean rotation error in both 3 and 6 DoF even when running on hardware with low computing capabilities (such as the Intel Atom N2800).

The next section describes the main processing modules of the DRL system. Section 4 details the testing platforms and configurations used, while section 5 provides an analysis of the test results. Section 6 finishes with the conclusions.

2. Brief presentation of the DRL system

The DRL system [9, 10] was implemented as a Robot Operating System (ROS) [11] package[1] and can perform 3 and 6 DoF pose estimations of mobile robot platforms. It was implemented as a set of modular C++ templated shared libraries in order to be reusable for other applications besides robot self-localization. It extensively uses the Point Cloud Library (PCL) [12] for point cloud preprocessing and registration and the OctoMap framework [13] for dynamic map update.

2.1. Processing pipeline configuration

Given the wide range of processing capabilities and sensor configurations that a mobile robot platform can have and the challenging environments in which they might operate, the DRL system can be fully configurable through yaml files in order to meet the specific needs of a given robot application while using the least amount of computational resources. To achieve these goals, it offers a flexible and dynamic pose estimation processing pipeline (brief overview shown in Figure 1) with a range of preprocessing algorithms along with three levels of point cloud registration. The first level is intended for the normal operation of the robot and can be configured for maximum efficiency and precision. The second level can be used for pose tracking recovery and can have algorithms and configurations able to recover from temporary tracking problems, such as partial sensor occlusions or unreliable odometry information. The third level of registration is able to estimate the initial pose of the robot when it starts operating or when it becomes lost in the environment. Besides pose estimation, the system can also incrementally build or update the environment map, allowing mobile robots to explore unknown areas and also leading to better path planning because the navigation system will have an updated view of its surroundings.

The DRL system was designed to operate with sensors capable of generating point clouds of the environment, and as such, it can directly use RGB-D and ToF cameras. For LIDARs, it offers a laser scan assembler that is able to merge measurements from several sensors using spherical linear interpolation (to reduce point cloud deformation). It also allows the usage of a circular

1 https://github.com/carlosmccosta/dynamic_robot_localization

buffer in order to register point clouds containing new measurements from the environment along with points that were previously registered (useful when using several sources of sensor data).

For fine tuning of the processing pipeline configuration, the self-localization system provides a detailed analysis of the estimated poses along with the computation time of each of its main processing stages. This allows to pinpoint the algorithms that require more processing resources, which can be valuable information when the system is running on platforms with very low processing capabilities. Moreover, these computation time logs also allow to identify which processing stages would benefit from a more suitable parameterization or a different algorithmic approach.

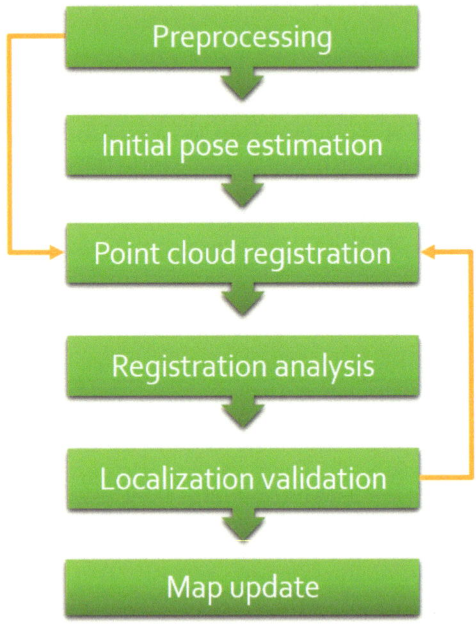

Figure 1. Localization system processing pipeline overview

2.2. Preprocessing

Preprocessing allows to adjust the level of detail of the reference or ambient point clouds and is also very effective in minimizing the impact of sensor noise (by using voxel grids, random sampling, outlier removal, surface reconstruction, sensor distance thresholds and color segmentation). Moreover, it can add information to the sensor data, such as line and surface normals and geometry descriptors, which are required to perform feature registration.

For fast deployment of mobile robots, the self-localization system supports two independent preprocessing pipelines: one for the reference point cloud (map of the environment) and another for the ambient and sensors point cloud. This allows the system to reuse maps from different sources and with different levels of detail while also giving some control over the time that will be required to perform the pose estimations, given that the amount of time required to register two point clouds decreases considerably when the number of points in both the reference and ambient point clouds are reduced.

2.3. Initial pose estimation

When a mobile robot platform starts its operation without a known initial pose or when it becomes lost in the environment, it requires registration methods capable of estimating the global pose in the known map. One category of algorithms that achieves these goals is the feature matching techniques. Algorithms in this category start with a keypoint detection phase in order to find geometric significant points in the environment and then describe each keypoint by analyzing its surrounding geometry, such as points and normals distribution. These keypoint descriptors (usually histograms) are then matched using a kd-tree and the best correspondences are found using a Random Sample Consensus (RANSAC) approach. With these point correspondences, the correction matrix can be computed and used to estimate the robot pose in the known map.

The keypoint selection phase is used mainly to reduce the computational resources required to perform the initial pose estimation, given that computing descriptors for every point would need a significant amount of processing time and wouldn't improve the feature registration significantly. However, for the geometric feature matching to be successful, the keypoint detector must be able to find the same keypoints when similar geometry is given, even when the sensor data is affected by noise and the point clouds have different level of detail.

The localization system currently supports the Scale Invariant Feature Transform (SIFT) [14] and the 3D Intrinsic Shape Signatures (ISS3D) [15] keypoints detectors. For keypoint description it can use the Point Feature Histogram (PFH) [16], the Fast Point Feature Histogram (FPFH) [17], the Signature of Histograms of Orientations (SHOT) [18], the Shape Context 3D (SC3D) [19], the Unique Shape Context (USC) [20] and the Ensemble of Shape Functions (ESF) [21].

2.4. Point cloud registration

Point cloud registration algorithms such as the Iterative Closest Point (ICP) [22] (with its known variations such as ICP point-to-point, ICP point-to-point non-linear, ICP point-to-plane and generalized ICP [23]) and also the Normal Distributions Transform (NDT) [6] can achieve accurate registration of point clouds by iteratively minimizing the alignment error, but they require the point clouds to be partially aligned in order to converge to a valid solution. As such, by combining the initial pose estimation methods with the point cloud registration algorithms, the DRL system can reliably and accurately estimate the robot pose.

2.5. Registration analysis and validation

Each pose estimation is followed by a post processing stage in which several registration metrics are computed and analyzed in order to determine if the point cloud registration was successful and the estimated pose is good enough to be considered valid. The first metrics computed are the inliers percentage and Root Mean Square Error (RMSE) (an inlier is a point in the sensor point cloud that has a point in the reference point cloud within a given distance radius). The second metric is the spatial angular distribution of both the inliers and outliers. This metric gives a measurement of confidence in the estimated pose and is based on the fact that when there are correctly registered points all around the robot, the confidence in the estimated pose is higher than when there is only correctly registered points in a small angular region of space. The last metrics are the translation and rotation corrections that were applied to the ambient point cloud in order to minimize the alignment error with the reference point cloud.

The configuration of these metrics thresholds allows to define what is considered a valid pose and may vary depending on the specific robot sensor configurations, the environment in which it will operate, the map resolution, and the required localization precision given the processing and capabilities of the robot.

These registration metrics also control when the self-localization system switches between the three point cloud registration modes. By default, the pose estimations are performed in the normal tracking mode. If the registration fails, the pose tracking recovery methods are activated. If the pose estimation continues to fail with new sensor data for a given period of time and a minimum number of pose estimations has been rejected, then the initial pose algorithms using feature registration are used to estimate the robot global pose and reset the tracking state.

2.6. Incremental map update

After successfully registering the 2D or 3D sensor data with the reference point cloud, the self-localization system can update the 2D or 3D map of the environment by integrating the fully registered point cloud or only the inliers or outliers. This allows to perform full integration of the sensor data when the environment is expected to change drastically or only inlier or outlier integration when small sections of the map need to be updated. By integrating only the outliers, the computation requirements are reduced considerably and also allow to avoid the degradation of the map by not integrating sensor data close to known areas (that could have been generated from CAD models or other accurate mapping systems). For 3D maps, surface reconstruction can be used to reduce the impact of sensor noise and to increase the quality and accuracy of the environment representation.

The self-localization system can also be paired with the OctoMap library in order to perform probabilistic integration of the sensor data and removal of missing ambient geometry.

3. Testing configurations

The DRL system was tested in three environments with three testing platforms in order to assess its accuracy and robustness when both the environment and its sensor configurations change.

http://dx.doi.org/10.5772/61258

The next sections provide a brief overview of the testing platforms used and the environments in which they were deployed.

3.1. Testing platforms

The self-localization system was tested with sensor data retrieved from two different mobile robot platforms (3 DoF tests) and also from a standalone Kinect sensor (6 DoF tests).

In order to allow the repetition and comparison of the tests in different computing platforms, the sensor data was recorded into rosbags and is available in the following repository[2].

The next sections provide a brief description of each of the testing platforms while Table 1 gives an overview of the sensors specifications.

3.1.1. Jarvis robot

The Jarvis robot (shown in Figure 2) is one of our autonomous ground vehicles and was equipped with a SICK NAV350 laser for self-localization (mounted about 2 m from the floor) and a SICK S3000 laser for collision avoidance (mounted about 0.1 m from the floor). It uses a tricycle locomotion system and had a ground truth provided by the SICK NAV350 system (relied on 6 lasers reflectors with 0.09 m of diameter).

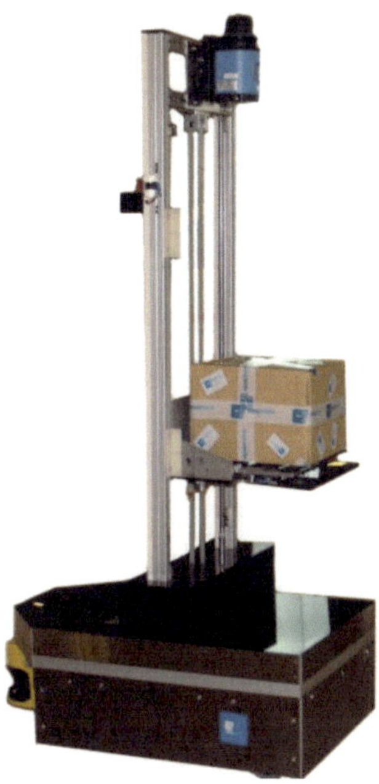

Figure 2. Jarvis testing platform

2 https://github.com/carlosmccosta/dynamic_robot_localization_tests

3.1.2. Pioneer 3-DX robot

The Pioneer 3-DX robot (shown in Figure 3 and presented in [24]) is a small autonomous vehicle equipped with a SICK LMS200 laser (mounted about 0.48 m from the floor) and a Kinect (mounted about 0.78 m from the floor). It uses a differential locomotion system and the ground truth was computed using 8 Raptor-E cameras.

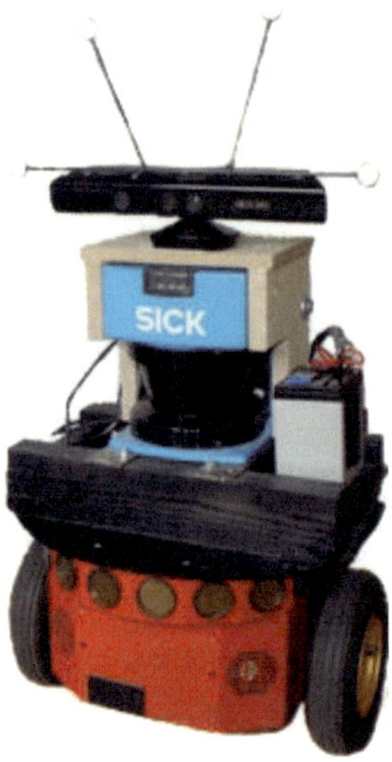

Figure 3. Pioneer 3-DX testing platform [24]

3.1.3. Kinect

The Kinect sensor (shown in Figure 4) is a structured light sensor capable of generating 3D colored point clouds of the environment at 30 Hz. It has a range of about 4 meters and has a vertical field of view of 43° and a horizontal field of view of 57°. It was moved within the testing area by a human operator and the ground truth was given by a Vicon motion tracking system.

Figure 4. Kinect sensor

Sensor	Number of measure-ments	Range (meters)	Field of view (degrees)	Scanning frequency (Hz)	Angular resolution (degrees)	Statistical error (millimeters)
SICK NAV350	1440	[0.5..250]	360H	8	0.25	15
SICK S3000	760	[0.1..49]	190H	8	0.25	150
SICK LMS200	180	[0.1..80]	180H	10	1.00	35
Kinect	307200	[0.8..4.0]	57H I 43V	30	0.09	10

Table 1. Sensors hardware specifications

3.2. Computing platforms

The accuracy and computational requirements of self-localization systems can change significantly depending on the environment, sensors used, and movement path of the robot. As such, in order to have representative results, the DRL system was tested in three computing platforms with very different processing capabilities and all running Ubuntu 12.04 along with ROS Hydro and PCL 1.7.

The next sections present a brief overview of each of these computing platforms while Table 2 provides a detailed description of the CPUs used.

3.2.1. High-performance laptop

The Clevo P370EM laptop is a 2012 high-performance laptop equipped with a quad-core Intel Core i7-3630QM CPU, 16 GB of DDR3 at 1600 MHz, and an NVidia GeForce GTX680M GPU.

3.2.2. Low-performance laptop

The Samsung 530U3C is a 2012 low-performance laptop equipped with a dual-core Intel Core i5-3317U CPU, 6 GB of DDR3 at 16000 MHz, and an Intel HD Graphics 4000 GPU.

3.2.3. Low-performance embedded PC

The 2012 low-performance embedded PC was installed in a Robotnik Guardian mobile robot platform and was equipped with a dual-core Intel Atom N2800 CPU, 2 GB of DDR3 at 1066 MHz, and an Intel Graphics Media Accelerator 3650 GPU.

3.2.4. Computing platforms comparison

When building mobile robots, autonomy can be a serious concern, and, as such, a CPU with very low power consumption and a passive or fanless cooling system (such as the Intel Atom N2800 with 6.5 W of Thermal Design Power (TDP)) might be enough to accomplish the desired tasks. When higher workloads are expected, a low-power mobile CPU (such as the Intel Core i5-3317U with a TDP of 17 W) might be a better choice. If power consumption (and cost) isn't an issue, then a high-performance CPU will be more than enough to run the DRL system in both 3 and 6 DoF along with the perception, planning, and decision modules.

Analyzing the benchmark results present in Table 2[3], it can be seen that the low-performance embedded PC had a CPU (Intel Atom N2800) about 10 times less capable than the high-performance laptop (Intel i7-3630QM), while the low-performance laptop (Intel i5-3317U) had only 2.5 times less computing capabilities. It should also be noted that the Intel Atom N2800 didn't have a level 3 cache, which could significantly improve the performance of the CPU when using applications that make intensive use of memory, which is the case of the DRL system. Moreover, the memory used in the embedded PC had about 50% less frequency (1066 MHz vs. 1600 MHz) and the Direct Memory Interface (DMI) had half the rate of the other two CPUs (2.5 GT/s vs. 5.0 GT/s), which causes cache misses to have high impact on overall CPU performance. Lastly, the Intel Atom had an operating frequency that was 1.29 times lower than the Intel i7-3630QM (and didn't have Intel Turbo Boost functionality).

	Intel Atom N2800	**Intel i5 3317U**	**Intel i7-3630QM**	**i7 / i5**	**i7 / Atom**	**i5 / Atom**
Number of cores	2	2	4	2.00	2.00	1.00
Minimum operating frequency (GHz)	1.867	1.7	2.4	1.41	1.29	0.91
Maximum operating frequency (GHz)	1.867	2.6	3.4	1.31	1.82	1.39
Direct Memory Interface (GT/s)	2.5	5.0	5.0	1.00	2.00	2.00
Level 1 instructions cache (KB)	2*32	2*32	4*32	2.00	2.00	1.00
Level 1 data cache (KB)	2*24	2*32	4*32	2.00	2.67	1.33
Level 2 cache (KB)	2*512	2*256	4*256	2.00	2.00	1.00
Level 3 cache (KB)	0	3072	6144	2.00	-	-
Number of transistors (billions)	0.176	0.634	1.4	2.21	7.95	3.60
Hyper-Threading	Yes	Yes	Yes	-	-	-
Manufacturing process (nm)	32	22	22	1.00	0.69	0.69
Thermal Design Power (watt)	6.5	17	45	2.65	6.92	2.62
Launch date	Q4-2011	Q2-2012	Q3-2012	-	-	-
Microarchitecture	Saltwell	Ivy Bridge	Ivy Bridge	-	-	-
Recommended Customer Price (dollars)	47	225	378	1.68	8.04	4.79
3D Mark 06 CPU score	965	2984	6392	2.14	6.62	3.09
Passmark score	636	3097	7766	2.51	12.21	4.87
GeekBench 32-bit score	1033	4039	10196	2.52	9.87	3.91
CINEbench 32-bit R10 Multi CPU score	1829	7337	18091	2.47	9.89	4.01

Table 2. Computing platforms CPU comparison

3 Retrieved from http://cpuboss.com/ and http://www.notebookcheck.net/

3.3. Testing environments

The self-localization system was tested in three different environments using the Jarvis robot in a RoboCup field, the Pioneer 3-DX robot in an industrial hall, and a Kinect sensor in a flying arena.

The next sections describe each of these testing environments.

3.3.1. Jarvis robot in a RoboCup field

The RoboCup field (shown in Figures 5 and 6) is on the right side of a large room with 20.5 m of length and 7.7 m of depth. It has two doors, several small windows, and three large glass openings into the hallway.

Several tests were performed in this environment, with the robot moving with speeds ranging from 0.05 m/s to 0.5 m/s. These tests were done with the robot following either a circular or a complex path containing linear and rotational movements with different speeds.

Figure 5. Jarvis testing environment (east side)

3.3.2. Pioneer 3-DX robot in an industrial hall

The industrial hall (shown in Figures 7 and 8 and presented in [24]) is a large room with long and high walls and several occluding objects in the middle.

The first test was done with the robot moving at 0.23 m/s, following a 360° path and without objects in the middle of the environment. The remaining tests were performed with the robot following different paths, with speeds ranging from 0.16 m/s to 0.26 m/s and with occluding objects in the middle.

Figure 6. Jarvis testing environment (west side)

Figure 7. Industrial hall overview [24]

Figure 8. Industrial hall with objects in the middle [24]

3.3.3. Kinect sensor in a flying arena

The flying arena (shown in Figure 9 and presented in [25]) is a large room where several objects with varying dimensions and shapes were added in order to test self-localization systems.

Three different tests were performed in this environment. In the first test, the Kinect was moved in a free-fly motion while the remaining two tests had movements with mainly translations or rotations.

4. Test results

This section presents the results achieved with the DRL system in each of the testing environments (shown from Table 3 to Table 8). They were performed in three different computing platforms with varying processing capabilities and with known initial poses. Each test has information about the conditions in which it was performed (duration, mean velocity of the robot and its path, and also the number of sensor measurements used), and the analysis of the self-localization system error (translation or rotation mean and standard deviation error and

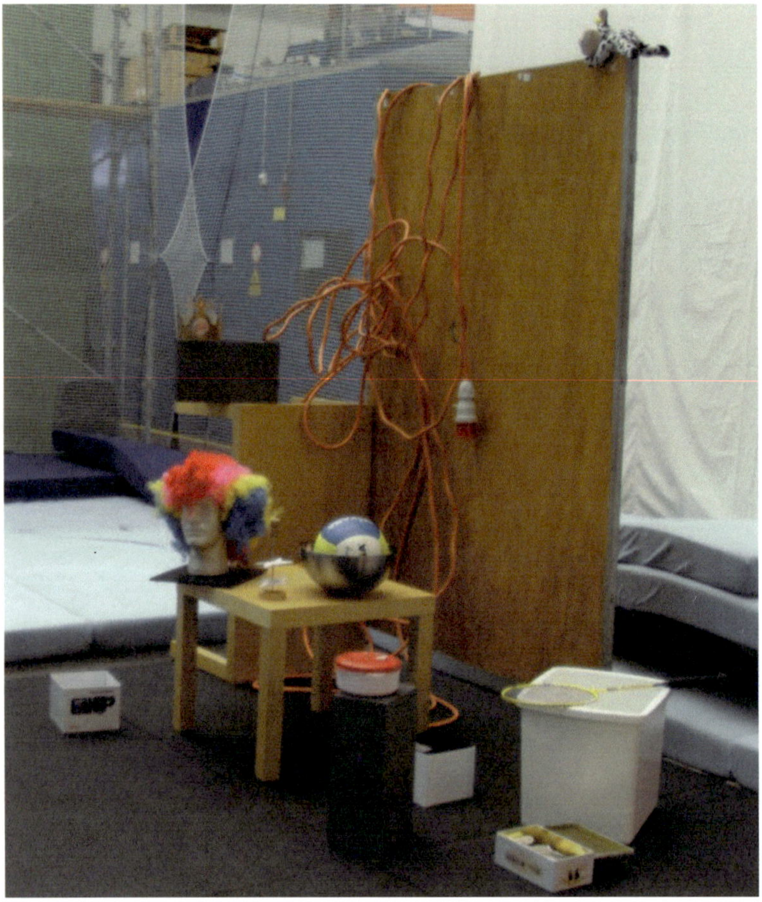

Figure 9. Kinect sensor flying arena [25]

also mean and standard deviation of the computation time used to perform the pose updates) along with the profiling of the computational resources required by the DRL system (using the collectl monitoring tool for the CPU and memory usage and the perf profiler for the remaining metrics).

4.1. 3 DoF test results

The 3 DoF tests relied on LIDAR data and used the 2D ICP (point-to-point) algorithm for pose tracking and 3D ICP (point-to-point) algorithm for tracking recovery (with a larger kd-tree search radius for finding point correspondences, higher limit for the number of iterations allowed in the point cloud registration and a higher convergence time limit).

The 3 DoF maps were generated with the self-localization system in SLAM mode (shown in Figures 10 and 15) and were manually corrected in order to improve the map accuracy and to keep only the most relevant geometry of the environment.

4.1.1. *Jarvis robot test results*

The 3 DoF tests performed with the Jarvis robot in the RoboCup field (presented in Tables 3 and 4 and from Figures 11 to 14) show that the DRL system was able to track the robot pose (on a map with 10 mm cell resolution) with high accuracy (4 to 12 mm of translation error and 0.4° to 0.7° of rotation error) on the three computing platforms (when using 500 points from sensor measurements). The localization system error (translation and rotation) was consistent across all testing platforms (deviation in translation error bellow 1 mm and in rotation error below 0.11° between tests, which shows the repeatability of the DRL system) and the mean computation time required to update the pose with new sensor data increased on the CPUs with lower processing capabilities. As expected, the computation time ratios between the different CPUs were similar to the benchmark performance ratios presented in Table 2. In this dataset, the Intel Atom had a mean computation time that was 7.69 times higher than the Intel i7, and the Intel i5 required only 1.35 more processing time than the Intel i7. Moreover, the standard deviation of the computation time also increased on the Intel Atom (6.43 times higher than the Intel i7) and in the Intel i5 (1.31 higher than the Intel i7). The CPU usage (percentage in tables are from 0–100% times the number of virtual cores, which in the case of a quad-core with Hyper-Threading would yield a range of [0-800]) followed the same trend as the computation time, while the memory required by the self-localization system remained consistent across all testing platforms (it is related to the map resolution and number of points in the reference or sensor point clouds).

Analyzing the data collected with the perf profiler, it can be seen that the Intel i7 had a mean operation frequency of 3.0 GHz, while the Intel i5 remained at 2.3 GHz (77% of the Intel i7 frequency) and the Intel Atom achieved only 1.82 GHz (62% of the Intel i7 frequency). Moreover, due to the lack of level 3 cache and smaller level 1 cache, the Intel Atom had smaller cache references rate (2.76 times less than the Intel i7) and significantly more cache misses (3.74 times more than the Core i7), which caused an increase of 37% in bus cycles in order to access the main memory. Besides cache misses, the Intel Atom also had much more branch misses (3.46 times more than the Intel i7), which led to less branch instructions per second (6.76 times less than the Intel i7). These memory access bottlenecks and branch mispredictions caused the mean number of instructions per cycle to drop to 0.32 (the Intel Atom has instruction pipelining and Hyper-Threading, which in ideal conditions allows it to process two instructions per clock cycle), which is an indicator that the CPU was spending a considerable amount of time waiting for data from the main memory (which can be several orders of magnitude slower than cache memory) or was wasting processing resources in branches that were not taken (resulting in the execution of instructions that won't be useful and can cause the flushing of the processing pipeline, resulting in even more waste of CPUs resources).

Given that most mobile robot platforms can be used to perform complex tasks in the environment besides localization and navigation, two more configurations were tried in each of the four tests (done with the Intel Atom) in order to assess the trade-offs between localization precision and computation time. This was achieved because the self-localization system can be tuned to specific requirements of accuracy (given the available sensors and computational resources). Looking at Tables 3 and 4, it can be seen that reducing the number of points used in the point cloud registration (retrieved from sensor data) to half (250) led to a reduction in computation time of 37% at the cost of increasing the localization translation error by 48%. Reducing the number of points to half again (125), resulted in a reduction of computation time of 62% and a translation error increase of 217% (in relation to the original test with 500 points). Despite the large increase in translation error when reducing the number of registration points, some robots may prefer a localization system with low computational requirements and a localization error close to a centimeter instead of having millimeter accuracy with moderate CPU usage.

Analyzing the remaining test results, it can be seen that the Intel i5 didn't suffer from the Intel Atom memory bottlenecks and had much better branch prediction (Intel Atom has a different micro-architecture than the Intel Core processors), having only a small increase in cache misses (14%) and a decrease in branch misses (4%) in relation to the Intel i7. Moreover, the number of instructions per cycle, along with bus cycles rate, was very similar to the Intel i7 (less than 2% difference), while the cache references rate and branch instructions rate decreased by 25% in relation to the Intel i7 CPU.

Comparing the overall results, the Intel Atom was capable of tracking the robot pose with high precision with moderate CPU usage. Nevertheless, for robots designed to perform complex tasks besides navigation, the Intel i5 is probably a better choice, given that the localization system was only using 10% of one of its CPU cores (instead of 50% of one of the Intel Atom cores).

Path	Test duration (seconds)	Velocity (meters / second)	CPU	Number of sensor points	Translation error (millimeters)		Rotation error (degrees)		Computation time for pose update (milliseconds)		Test nº
					Mean	Standard deviation	Mean	Standard deviation	Mean	Standard deviation	
Circular	290	0.05	Intel Atom N2800	125	12.16	8.24	0.71	0.19	18.81	6.26	1.01
				250	5.69	3.40	0.60	0.10	31.20	11.11	1.02
				500	3.83	2.33	0.56	0.09	49.25	17.09	1.03
			Intel i5 3317U	500	4.15	2.38	0.56	0.09	9.05	3.24	1.04
			Intel i7-3630QM	500	3.97	2.38	0.56	0.09	6.57	2.41	1.05
	25.5	0.5	Intel Atom N2800	125	14.31	9.04	0.61	0.46	23.71	6.75	1.06
				250	12.40	8.53	0.57	0.43	43.55	12.83	1.07
				500	12.20	8.72	0.54	0.41	79.92	17.01	1.08
			Intel i5 3317U	500	11.89	8.66	0.54	0.41	14.80	4.61	1.09

Path	Test duration (seconds)	Velocity (meters / second)	CPU	Number of sensor points	Translation error (millimeters)		Rotation error (degrees)		Computation time for pose update (milliseconds)		Test nº
					Mean	Standard deviation	Mean	Standard deviation	Mean	Standard deviation	
			Intel i7-3630QM	500	12.06	8.53	0.55	0.41	10.95	3.51	1.10
Complex	498	0.05	Intel Atom N2800	125	12.60	7.46	0.53	0.16	20.08	7.09	1.11
				250	7.82	4.07	0.49	0.10	34.64	13.10	1.12
				500	5.07	2.71	0.47	0.08	57.78	18.50	1.13
			Intel i5 3317U	500	5.13	2.75	0.45	0.08	9.60	3.39	1.14
			Intel i7-3630QM	500	4.86	2.67	0.46	0.07	7.20	2.66	1.15
	392	0.5-0.3-0.5-0.1	Intel Atom N2800	125	12.86	7.42	0.49	0.14	19.89	6.74	1.16
				250	8.52	4.84	0.45	0.12	35.73	13.48	1.17
				500	6.79	4.04	0.42	0.10	60.79	20.17	1.18
			Intel i5 3317U	500	6.74	3.92	0.42	0.10	10.35	3.88	1.19
			Intel i7-3630QM	500	6.74	3.92	0.42	0.10	7.66	2.95	1.20

Table 3. Self-localization test results performed with the Jarvis robot in the RoboCup field

Path	CPU usage (% 0..100 * nº cores)		Memory usage (MB)		CPU cycles (GHz)	Instruc-tions per cycle	Cache referen-ces (M/sec)	Cache misses (% of all cache references)	Branch instruc-tions (M/sec)	Branch misses (% of all branches)	Bus cycles (M/sec)	Test nº
	Mean	Standard deviation	Mean	Standard deviation								
Circular	33.35	14.23	65.46	5.24	1.83	0.22	15.18	27.66	80.43	7.24	130.24	1.01
	41.23	17.57	65.48	5.23	1.83	0.27	13.72	25.64	96.93	6.48	130.86	1.02
	52.10	23.61	65.86	5.34	1.83	0.31	12.32	23.76	109.23	5.95	131.23	1.03
	10.64	8.69	67.53	4.60	2.34	1.31	25.02	7.01	554.37	1.66	94.10	1.04
	7.57	7.59	87.22	4.86	3.02	1.27	36.75	5.47	768.84	1.69	96.66	1.05
	36.61	28.01	59.45	9.68	1.83	0.28	14.61	22.59	99.64	5.86	130.66	1.06

Path	CPU usage (% 0..100 * nº cores)		Memory usage (MB)		CPU cycles (GHz)	Instructions per cycle	Cache references (M/sec)	Cache misses (% of all cache references)	Branch instructions (M/sec)	Branch misses (% of all branches)	Bus cycles (M/sec)	Test nº
	Mean	Standard deviation	Mean	Standard deviation								
	48.55	29.45	59.73	9.91	1.83	0.31	13.14	21.71	107.96	5.71	131.03	1.07
	63.82	37.88	59.82	9.94	1.84	0.35	11.52	20.25	122.21	5.41	131.01	1.08
	12.76	15.50	59.65	4.01	2.20	1.42	23.46	8.25	598.90	1.55	93.53	1.09
	8.86	13.40	78.82	3.22	2.84	1.46	25.72	8.44	788.61	1.47	94.57	1.10
Complex	32.71	13.86	70.32	7.59	1.82	0.22	15.22	28.13	77.47	7.43	130.55	1.11
	41.18	18.86	70.41	7.56	1.83	0.27	13.66	25.84	93.93	6.59	130.98	1.12
	55.16	25.20	70.47	7.58	1.83	0.32	12.26	22.75	108.13	5.96	131.30	1.13
	10.32	8.31	72.72	8.14	2.30	1.37	25.39	6.51	596.54	1.62	95.74	1.14
	8.28	7.40	91.27	7.59	3.04	1.30	37.44	5.12	741.56	1.82	96.07	1.15
	31.69	14.18	67.72	6.23	1.82	0.21	15.44	28.26	75.27	7.58	130.55	1.16
	39.78	19.91	67.97	6.37	1.83	0.26	14.01	25.55	91.91	6.66	130.93	1.17
	53.06	27.91	67.84	6.49	1.83	0.31	12.47	22.85	107.72	5.98	131.19	1.18
	10.38	8.76	67.85	5.93	2.33	1.32	26.37	6.34	589.79	1.63	95.57	1.19
	8.09	7.82	88.94	6.03	3.04	1.28	37.05	6.05	725.08	1.78	95.70	1.20

Table 4. Profiling of self-localization tests performed with the Jarvis robot in the RoboCup field

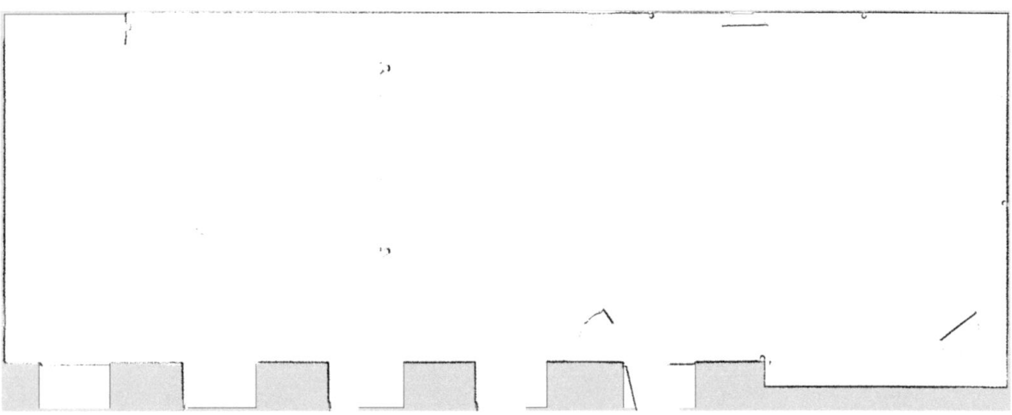

Figure 10. Map of the RoboCup testing environment using the DRL system in SLAM mode

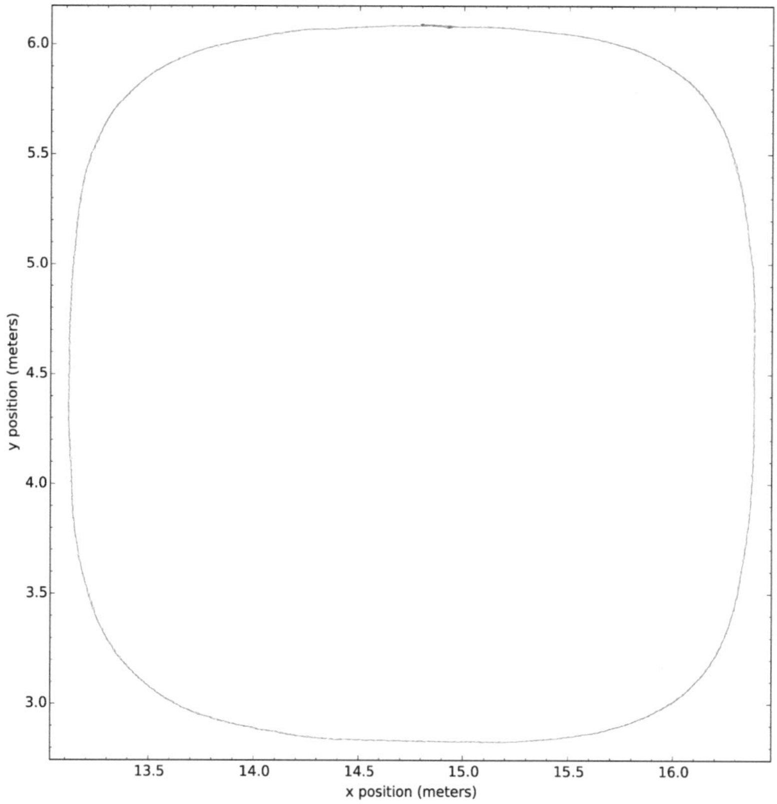

Figure 11. Poses estimated by the DRL (blue) and the ground truth (green) systems for test 1.03

4.1.2. Pioneer 3-DX robot results

The 3 DoF tests performed with the Pioneer robot in the industrial hall aimed to assess the robustness of the localization system on more challenging environments. In this dataset, the LIDAR sensor (SICK LMS200) had half the field of view of the sensor used in the previous dataset (SICK NAV350), and the angular resolution was 4 times lower, which resulted on 180 points for each laser scan (the previous dataset had 1,440 points per laser scan spread across 360°). Moreover, the resolution of the map (shown in Figure 15) was 2.5 times lower (25 mm cell resolution) and the laser scan measurements were limited to 10 m on the tests that had no objects in the middle of the environment (test 2.01, 2.02, and 2.03) and 6 m on the remaining tests (from test 2.04 to 2.12 in which there were objects in the middle of the environment that occluded the walls).

Analyzing Table 5 and 6 and Figures 16 to 19, it can be concluded that decreasing the field of view and reducing the map resolution resulted in a 5-time increase of the translation error and a 9-time increase in rotation error (it should be noted that the translation error remained below the map resolution, which is compelling evidence that the algorithms used can perform point

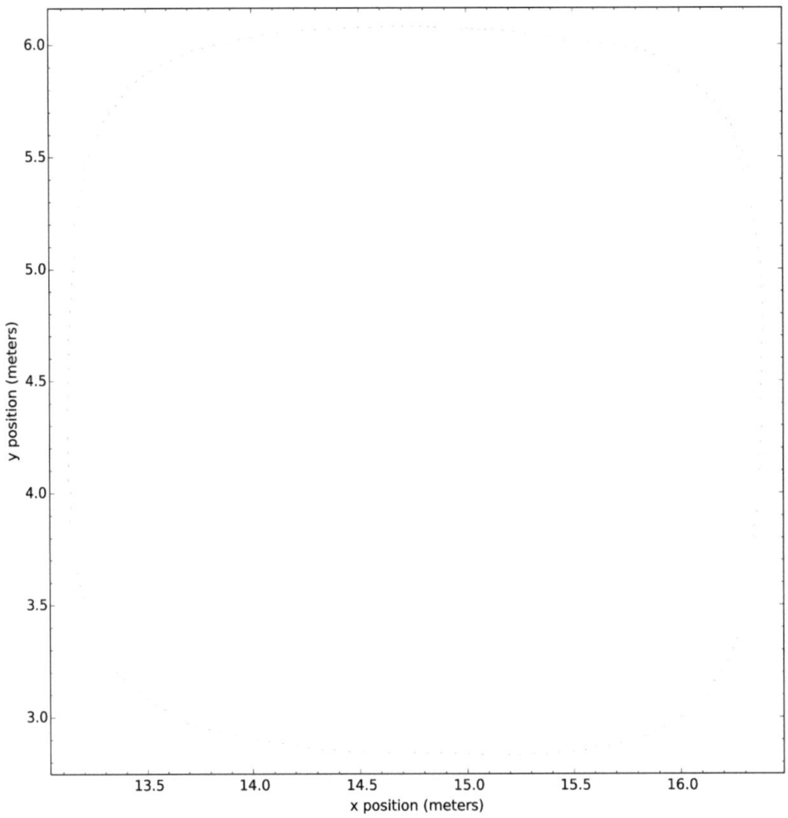

Figure 12. Poses estimated by the DRL (blue) and the ground truth (green) systems for test 1.08

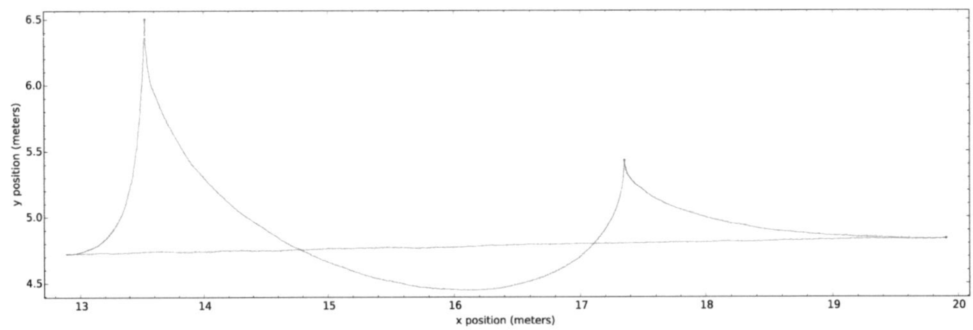

Figure 13. Poses estimated by the DRL (blue) and the ground truth (green) systems for test 1.13

cloud registration with high accuracy). The computation time was reduced about 40% on the Intel Core i7 and i5 (in relation to the previous dataset), and was mostly related to the fact that the number of points in both the reference and sensor point clouds was reduced by 33%. However, the computation time on the Intel Atom remained similar on both 3 DoF datasets

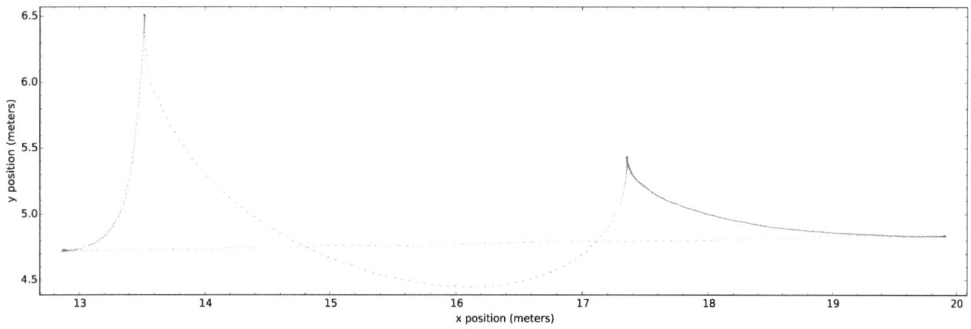

Figure 14. Poses estimated by the DRL (blue) and the ground truth (green) systems for test 1.18

because the percentage of branch misses increased and the mean number of instructions per cycle was reduced (from 0.32 to 0.24).

Comparing the results across the three computational platforms, it can be seen that both translation and rotation error distributions remained consistent across the three computing platforms and had similar results (about 2% change). The Intel i5 had very similar profiling results in relation to the Intel i7, while the Intel Atom required much more time to compute the results (due to its higher percentage of cache and branch misses). As such, similar to the previous dataset, the Intel i5 would be more suitable to run the DRL system since it doesn't suffer from the memory access bottlenecks and was able run the localization system with low CPU usage while requiring much less power than the Intel i7 (17 W vs. 45 W TDP).

Path	Test duration (seconds)	Velocity (meters / second)	CPU	Number of sensor points	Translation error (millimeters)		Rotation error (degrees)		Computation time for pose update (milliseconds)		Test nº
					Mean	Standard deviation	Mean	Standard deviation	Mean	Standard deviation	
360	72	0.23	Intel Atom N2800	180	20.32	11.09	5.71	0.66	59.23	22.76	2.01
			Intel i5 3317U		19.86	10.20	5.71	0.66	6.67	2.31	2.02
			Intel i7-3630QM		19.86	10.19	5.71	0.66	4.69	1.50	2.03
Slam 1	155	0.26	Intel Atom N2800		22.49	12.46	5.44	0.68	51.98	25.26	2.04
			Intel i5 3317U		22.34	12.19	5.43	0.68	5.74	2.53	2.05
			Intel i7-3630QM		22.35	12.20	5.43	0.68	3.92	1.66	2.06
Slam 2	115	0.19	Intel Atom N2800		19.67	13.09	5.52	0.82	54.36	24.60	2.07
			Intel i5 3317U		19.50	12.94	5.53	0.82	6.08	2.55	2.08
			Intel i7-3630QM		19.48	12.91	5.53	0.82	4.16	1.66	2.09

Path	Test duration (seconds)	Velocity (meters / second)	CPU	Number of sensor points	Translation error (millimeters)		Rotation error (degrees)		Computation time for pose update (milliseconds)		Test nº
					Mean	Standard deviation	Mean	Standard deviation	Mean	Standard deviation	
Slam 3	112	0.16	Intel Atom N2800		17.93	9.14	5.46	0.72	49.41	24.59	2.10
					17.93	9.14	5.46	0.72	49.41	24.59	2.10
	112	0.16	Intel i5 3317U		17.68	8.93	5.45	0.73	5.54	2.48	2.11
			Intel i7-3630QM		17.68	8.93	5.46	0.73	3.87	1.67	2.12

Table 5. Self-localization test results performed with the Pioneer robot in the industrial hall

Path	CPU usage (% 0..100 * nº cores)		Memory usage (MB)		CPU cycles (GHz)	Instruc-tions per cycle	Cache references (M/sec)	Cache misses (% of all cache references)	Branch instructions (M/sec)	Branch misses (% of all branches)	Bus cycles (M/sec)	Test nº
	Mean	Standard deviation	Mean	Standard deviation								
360	62.55	28.56	61.94	4.07	1.82	0.24	16.12	19.90	86.44	7.54	130.48	2.01
	14.06	10.99	64.08	4.12	2.24	1.03	33.19	5.96	464.40	1.82	95.02	2.02
	11.79	10.37	85.02	4.08	2.93	0.96	51.91	1.83	586.10	2.03	94.95	2.03
Slam 1	61.84	23.87	67.99	6.85	1.82	0.23	16.19	21.24	82.17	7.77	130.48	2.04
	13.98	9.82	71.11	7.68	2.25	0.93	38.20	4.75	416.97	2.02	94.66	2.05
	11.00	8.69	94.15	7.68	2.99	0.95	59.02	1.71	567.99	2.10	95.57	2.06
Slam 2	61.35	25.76	65.02	5.53	1.83	0.23	16.31	20.40	84.66	7.62	130.23	2.07
	14.13	10.18	67.86	5.98	2.28	0.64	35.70	4.73	437.17	1.89	94.28	2.08
	11.36	9.62	89.02	6.13	3.00	0.92	57.14	1.72	567.34	2.00	94.03	2.09
Slam 3	60.24	25.44	65.16	5.32	1.82	0.23	16.28	21.17	83.64	7.65	130.58	2.10
	13.20	10.53	67.81	5.96	2.14	0.96	35.90	5.50	433.83	1.92	95.01	2.11
	11.00	10.35	88.74	5.84	2.94	0.94	57.45	1.94	579.68	2.04	94.27	2.12

Table 6. Profiling of self-localization tests performed with the Pioneer robot in the industrial hall

4.2. 6 DoF Kinect sensor results

The 6 DoF tests performed with a standalone Kinect sensor in the flying arena (shown in Tables 7 and 8 and from Figure 22 to Figure 27) were meant to evaluate the accuracy and robustness of the self-localization system when moving the sensor in 6 DoF and following challenging paths in cluttered environments.

The 3D map (shown in Figures 20 and 21) was built using only the sensor data from the free-fly test (3.03) with the DRL system in SLAM mode and used surface reconstruction to automatically reduce the impact of sensor noise and improve its accuracy. Given the different paths

Figure 15. Map of the industrial hall testing environment using the DRL system in SLAM mode

that the Kinect had in each test, there were times in which the sensor field of view was outside of the known map (mainly in the test with rotations). The map was not extended to contain all the areas in the three tests in order to evaluate the robustness of the system against sensor occlusion or malfunctions (which are problems similar to the sensor seeing unknown areas). To allow real-time processing of the Kinect sensor data and keep the CPU usage under acceptable thresholds, the map was downsampled with a voxel grid of 20-mm cell size in both the Intel i7 and Intel i5 tests and a voxel grid of 50-mm cell size on the Intel Atom tests (to allow higher pose update rate).

Unlike the previous 3 DoF tests, estimating a pose in 6 DoF requires a reference point cloud with much more points (in order to represent the environment with the required level of detail), which leads to slower neighbor point searches (because the kd-tree used to perform the neighbor point searches will have more levels and it will take longer to retrieve the desired points). Given that point cloud registration algorithms make heavy use of point searches, the cumulative effect of increasing the search time led to the significant increase of the computation time and CPU usage seen in Tables 7 and 8 (when compared with the 3 DoF tests). Moreover, the more intensive use of memory caused the percentage of cache misses to increase signifi-

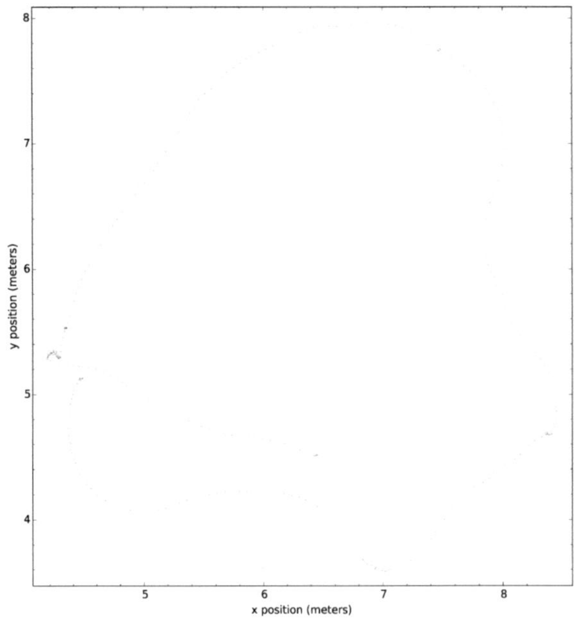

Figure 16. Poses estimated by the DRL (blue) and the ground truth (green) systems for test 2.01

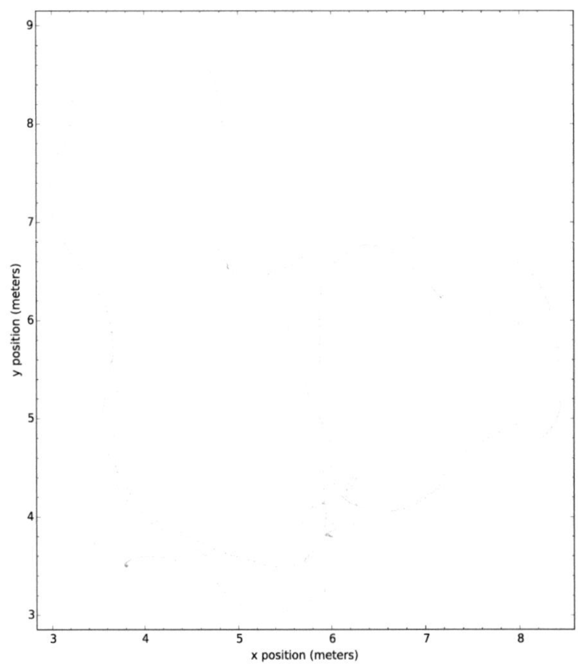

Figure 17. Poses estimated by the DRL (blue) and the ground truth (green) systems for test 2.04

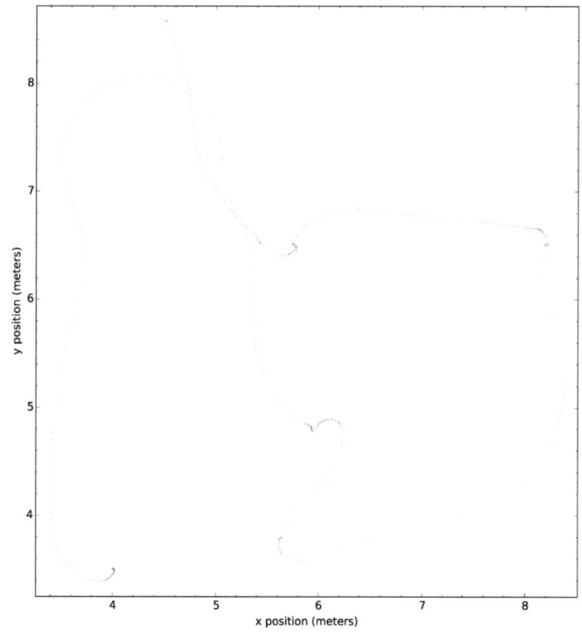

Figure 18. Poses estimated by the DRL (blue) and the ground truth (green) systems for test 2.07

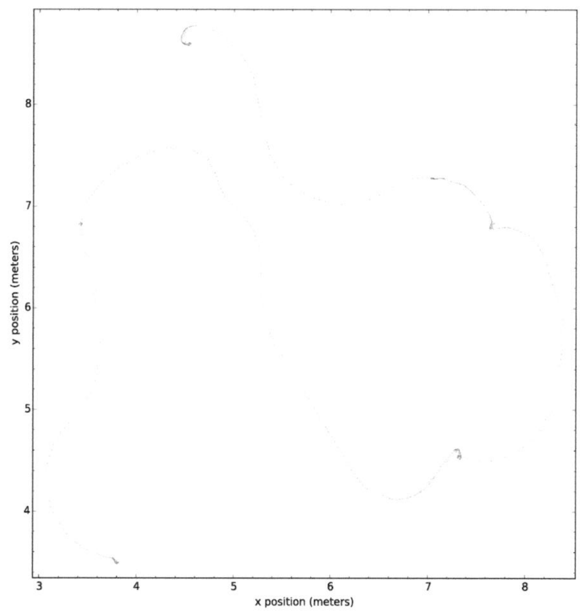

Figure 19. Poses estimated by the DRL (blue) and the ground truth (green) systems for test 2.10

cantly even on the Intel i5 and i7 (from 5–7% to 10–20%). This computational time increase was clearly seen in the test with mainly rotations in which the pose recovery systems were being activated when the sensor had its field of view outside the map. These recovery systems relied on the ICP point-to-plane (with a larger point search radius and higher convergence time limit) in order to be able to reset the tracking state (the normal tracking systems were using ICP point-to-point algorithm). Moreover, the maximum number of points used in this test was increased to make the pose recovery more robust, which also contributed to the increase in computation time (when compared with the other two 6 DoF tests).

Path	Test duration (seconds)	Velocity (meters / second)	CPU	Number of sensor points	Translation error (millimeters)		Rotation error (degrees)		Computation time for pose update (milliseconds)		Test nº
					Mean	Standard deviation	Mean	Standard deviation	Mean	Standard deviation	
Free fly	16.9	0.30	Intel Atom N2800	150	26.90	13.59	3.22	0.88	53.79	20.42	3.01
			Intel i5 3317U	425	19.16	10.19	3.07	0.67	35.03	9.75	3.02
			Intel i7-3630QM	425	19.44	9.63	3.07	0.64	25.94	7.15	3.03
Transla-tions	32.1	0.20	Intel Atom N2800	150	24.09	12.93	2.73	0.71	48.77	16.39	3.04
			Intel i5 3317U	425	16.89	9.83	2.75	0.57	33.58	9.30	3.05
			Intel i7-3630QM	425	16.18	7.84	2.76	0.57	24.83	6.61	3.06
Rotations	33.4	0.10	Intel Atom N2800	150	31.11	17.46	2.96	1.11	58.35	14.49	3.07
			Intel i5 3317U	425	18.67	12.61	2.53	0.96	62.85	25.40	3.08
			Intel i7-3630QM	750	18.82	12.30	2.55	0.92	41.51	17.39	3.09

Table 7. Self-localization test results performed with the Kinect sensor in the flying arena

Overall, the DRL system was able to track the Kinect pose with a mean translation error of 18 mm and a mean rotation error of 2.8° in the Intel i7 and i5 CPUs (it should be noted that the map resolution was 20 mm). For the Intel Atom, the number of points used in the registration (retrieved from the Kinect sensor) needed to be reduced by 60% in relation to the other CPUs, because the excessive computation time was leading to very high CPU usage and causing the localization system to ignore a significant amount of Kinect sensor scans. Adjusting the number of points increased the pose estimation error but allowed the Intel Atom to estimate the Kinect pose more often (even though it couldn´t update it at the same rate as the other CPUs, as can be seen from Figures 22 to 27). Nevertheless, the Intel Atom was able to track the Kinect pose with 20–30 mm of mean translation error and 2–3° of mean rotation error with an acceptable update rate (30–50% less update rate than the Intel i5 and i7).

Figure 20. Overview of the map of the flying arena environment using the DRL system in SLAM mode

Path	CPU usage (% 0..100 * nº cores)		Memory usage (MB)		CPU cycles (GHz)	Instruc-tionsper cycle	Cache references (M/sec)	Cache misses (% of all cache references)	Branch instruc-tions (M/sec)	Branch misses (% of all branches)	Bus cycles (M/sec)	Test nº
	Mean	Standard deviation	Mean	Standard deviation								
Free fly	81.60	53.07	99.11	11.08	1.84	0.35	11.00	18.51	126.21	5.35	131.62	3.01
	54.42	50.29	107.18	5.28	2.39	1.61	11.33	18.75	734.72	1.96	95.89	3.02
	48.60	41.72	125.92	4.03	3.14	1.65	14.51	11.35	965.25	1.98	97.50	3.03
Transla-tions	84.91	52.01	100.22	8.45	1.84	0.34	10.88	18.12	122.53	5.34	131.73	3.04
	74.40	43.35	108.12	4.85	2.40	1.61	11.73	17.19	720.56	1.97	96.77	3.05
	57.82	36.14	126.76	3.78	3.15	1.61	15.74	11.99	953.30	1.99	98.34	3.06
Rotations	97.25	42.98	100.09	9.83	1.84	0.35	10.94	18.43	121.57	5.66	131.71	3.07
	78.67	44.90	106.99	4.96	2.33	1.64	9.73	18.91	690.02	2.05	96.67	3.08
	76.16	43.72	129.64	4.08	3.15	1.69	13.13	9.93	1000.23	1.98	98.35	3.09

Table 8. Profiling of self-localization tests performed with the Kinect sensor in the flying arena

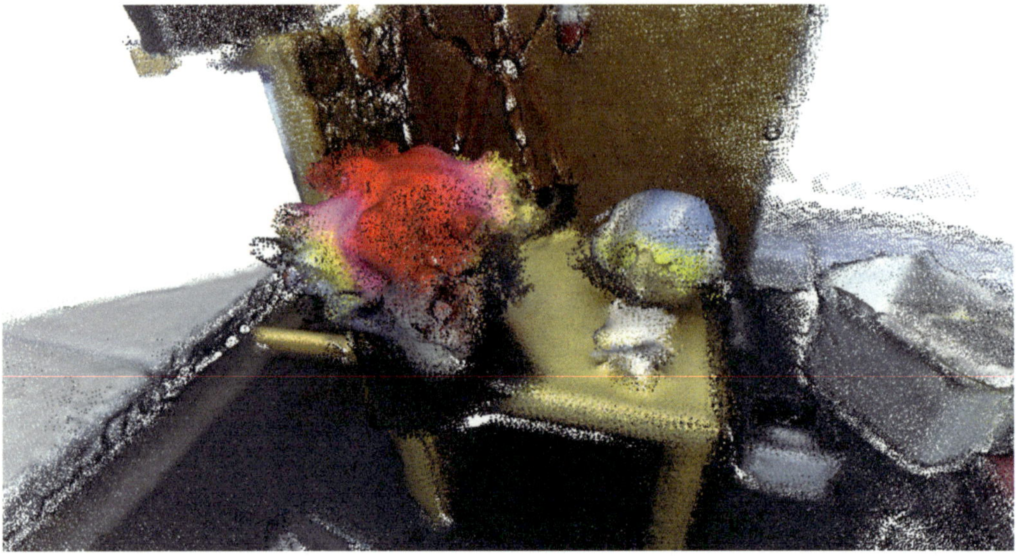

Figure 21. Map of the flying arena using the self-localization system in SLAM mode

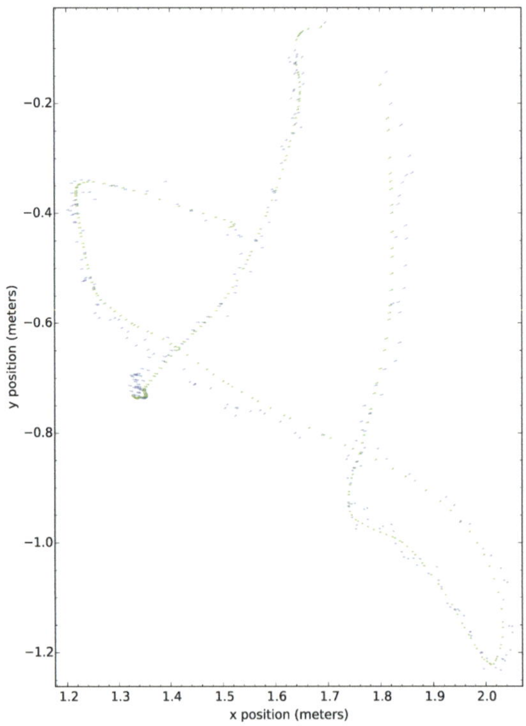

Figure 22. Poses estimated by the DRL (blue) and the ground truth (green) systems for test 3.01

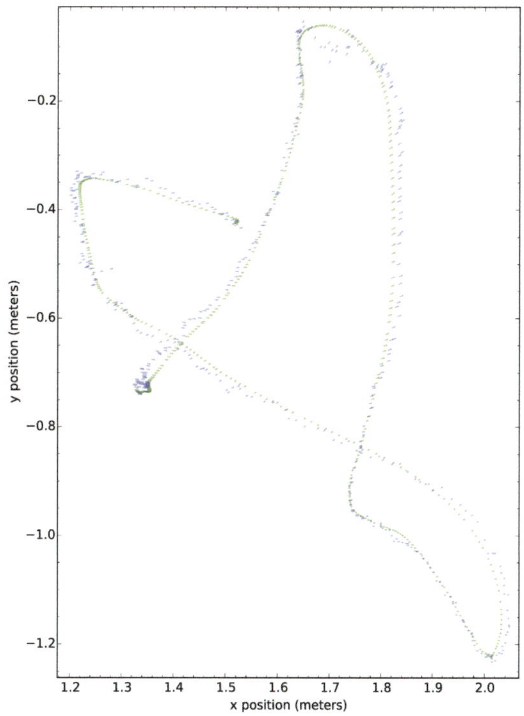

Figure 23. Poses estimated by the DRL (blue) and the ground truth (green) systems for test 3.02

5. Global analysis of test results

Analyzing the test results shown from Tables 3 to 8 and Figures 28 to 31 (with the testing conditions presented in Table 9), it can be concluded that the DRL system can achieve very accurate 3 DoF pose tracking (mean translation error below 1 cm and mean rotation error below 1°) when the sensor data provided has a wide field of view (360°), a high angular resolution (0.25°), and a long range (250 m). These results (configurations j1 to j4) hold even when the Jarvis testing platform was moving at varying speeds (from 0.05 to 0.5 m/s) and with dynamic objects in the environment.

When the DRL system was tested under more challenging conditions (configurations p1 to p4 in the industrial hall), with the sensor data having a reduced field of view (180°), low angular resolution (1°), and short range (6 m), the DRL system still managed to track the robot 3 DoF pose with translation error below the map resolution (mean translation error below 2 cm and mean rotation error below 6°).

The DRL system was also able to track the Kinect 6 DoF pose (configurations k1 to k3) with less than 2 cm of mean translation error and less than 3° of mean rotation error, even when the sensor was aimed at unknown areas, which shows the robustness of the DRL system against temporary sensor problems, which happen more often on sensors with narrow field of view.

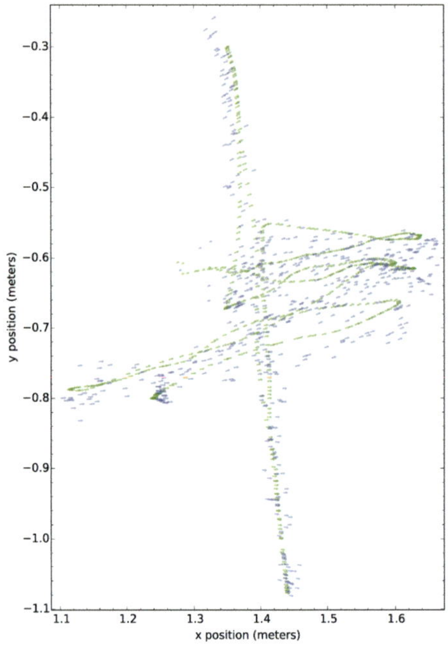

Figure 24. Poses estimated by the DRL (blue) and the ground truth (green) systems for test 3.04

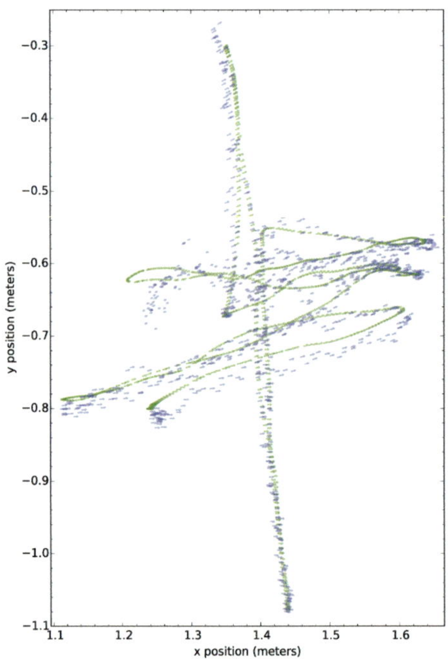

Figure 25. Poses estimated by the DRL (blue) and the ground truth (green) systems for test 3.05

http://dx.doi.org/10.5772/61258

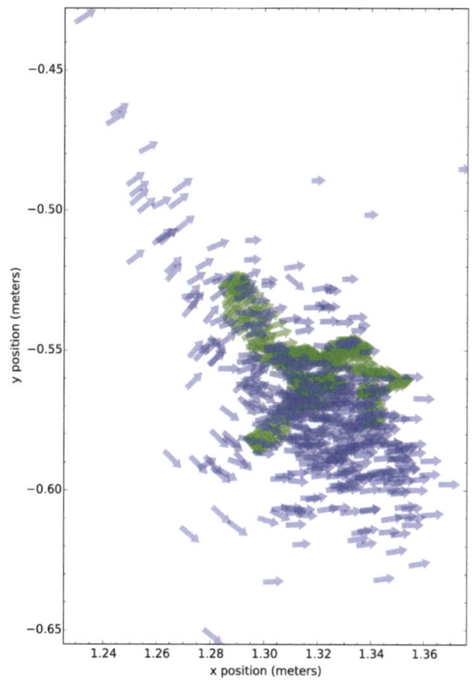

Figure 26. Poses estimated by the DRL (blue) and the ground truth (green) systems for test 3.07

Overall, the DRL system was able to track the 3/6 DoF pose of the testing platforms (equipped with significantly different sensors) with high accuracy and reliability and overcame temporary sensor occlusions and dynamic objects in the environment.

Comparing the profiling results of all tests, it can be concluded that using more points in the reference or sensor point clouds leads to higher computation times (for each pose update) and higher CPU usage. This can be seen in the 6 DoF test results, which had a 5-time increase in the mean computation time and CPU usage when compared to the 3 DoF tests. As such, the 6 DoF pose estimation capabilities of the DRL system should only be used when the mobile platforms are not moving on planar environments.

The DRL system can perform 3/6 DoF pose estimation with CPUs with low power consumption and cost, such as the Intel Atom N2800. However, the standard deviation of the pose estimation, along with the resources required, will grow when compared with better CPUs. Moreover, these low-consumption CPUs may have memory access bottlenecks and given their limited processing capabilities, they may require the tuning of the DRL configurations in order to be able to perform pose estimations with acceptable update rate (which is why Figures 30 and 31 show a similar 3/6 DoF computation time and CPU usage on the tests performed with the Intel Atom N2800). This reconfiguration allowed the DRL system to update the Kinect pose more often at the cost of higher translation and rotation error. Nevertheless, this shows that the DRL system can be used on a wide range of mobile robot platforms with varying computational capabilities.

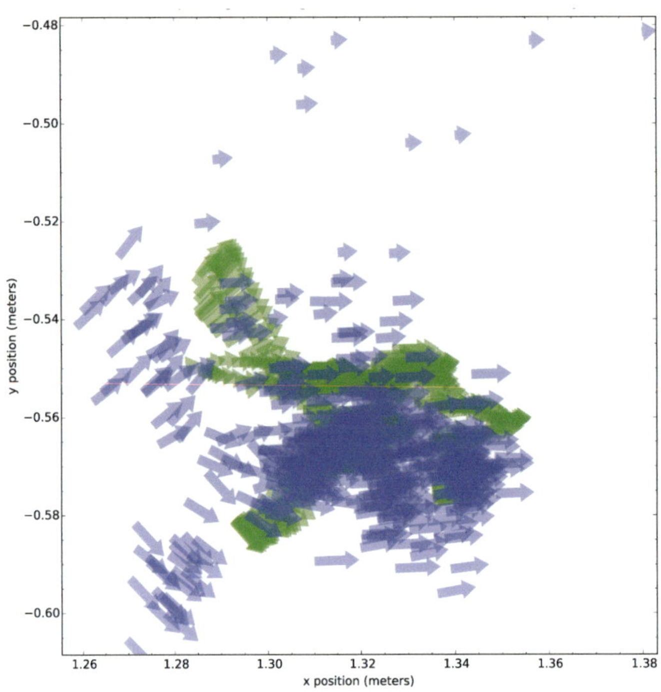

Figure 27. Poses estimated by the DRL (blue) and the ground truth (green) systems for test 3.08

Testing configurations	Sensor platforms	Movement path	Test duration (s)	Mean velocity (m/2)	Map resolution (mm)	Sensor range (m)	Filters	
							Voxel grid (m)	Random Sample
j1	Jarvis	Circular	290	0.05	10	250	0.02	500
j2			25.5	0.5				
j3		Complex	498	0.05				
j4			392	0.5-0.3-0.5-0.1				
p1	Pioneer	360	72	0.23	25	10	-	-
p2		Slam 1	155	0.26		6	-	-
p3		Slam 2	115	0.19			-	-
p4		Slam 3	112	0.16			-	-
k1	Kinect	Free fly	16.9	0.3	20	3.5	Atom -> 0.05 I i5/i7 -> 0.02	Atom -> 175 I i5/i7 -> 425
K2		Translations	32.1	0.2				Atom -> 175 I i5/i7 -> 425
K3		Rotations	33.4	0.1				Atom -> 175 I i5/i7 -> 750

Table 9. Testing configurations shown from Figure 28 to 31

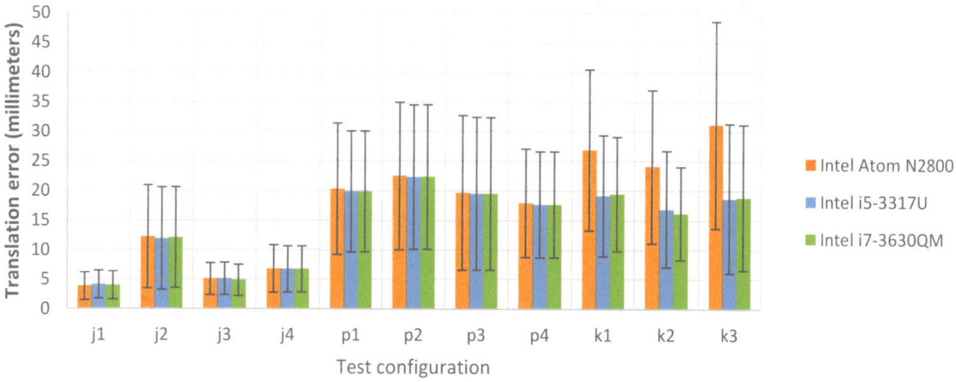

Figure 28. Translation error across all tests (testing configurations shown in Table 9)

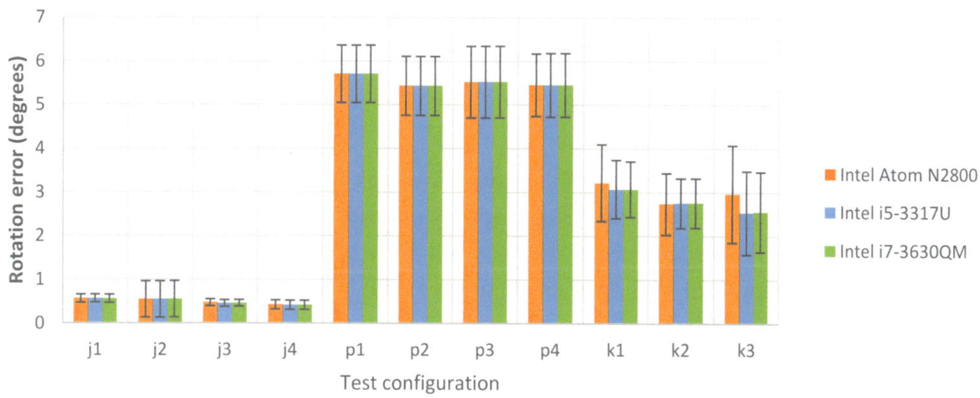

Figure 29. Rotation error across all tests (testing configurations shown in Table 9)

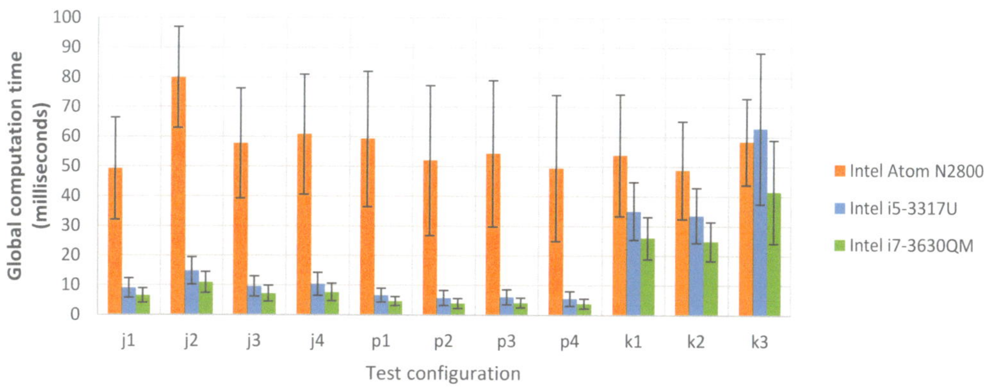

Figure 30. Computation time across all tests (testing configurations shown in Table 9)

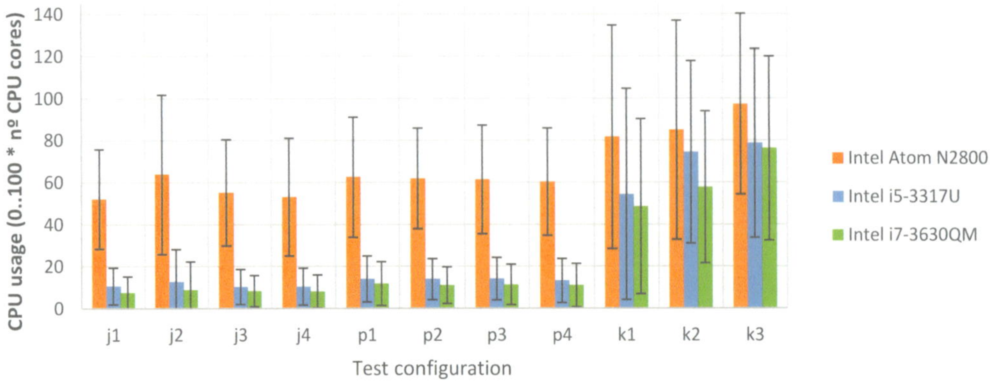

Figure 31. CPU usage across all tests (testing configurations shown in Table 9)

6. Conclusions

Mobile robot platforms require efficient software systems in order to perform their desired tasks without needing expensive and high power consumption hardware. Given the wide range of hardware and sensor configurations and the set of tasks that a mobile robot can perform, this article presented a detailed analysis of the DRL system performance in tests running on three different computing platforms equipped with CPUs ranging from low to high end processing capabilities.

The DRL system was tested in challenging environments and was able to perform high accuracy pose estimation with mean translation error between 5 and 30 mm and mean rotation error between 0.4° and 5° in both 3 and 6 DoF. The trade-offs between pose estimation accuracy and computing resources required can be tuned to the specific needs of the tasks performed by the robot, allowing efficient use of the localization system on low computing power mobile robot platforms.

For the presented tests, and with some configuration fine tuning, the Atom N2800 CPU was able to estimate the 6DoF pose with about 30 mm/3° in 60 ms at over 90% CPU load. The other superior CPUs, Intel Core i5 and i7, were able to estimate 6 DoF poses with about 20 mm/3° in 30 ms at 80% CPU load.

Moreover, several sensors can be used simultaneously in order to increase the field of view seen by the localization system, allowing more accurate and stable estimation of the robot's pose. Besides pose tracking, the self-localization system can also perform initial pose estimation when the robot starts its operation or when it becomes lost in the environment. It can also incrementally build a map of its surroundings with probabilistic integration and removal of geometry and perform surface reconstruction to minimize the impact of sensor noise.

The robust and high accuracy pose tracking capabilities of the DRL system in conjunction with the global pose estimation and mapping modules allow the fast deployment of a wide range of mobile robot platforms in cluttered and dynamic environments.

Acknowledgements

The authors would like to thank everyone involved in the CARLoS Project. This project has received funding from the European Commission Seventh Framework Programme for Research and Technological Development under the grant agreement number 606363, and from the national project "NORTE-07-0124-FEDER-000060".

Author details

Carlos M. Costa, Héber M. Sobreira, Armando J. Sousa* and Germano Veiga

*Address all correspondence to: asousa@fe.up.pt

INESC TEC (formerly INESC Porto) and Faculty of Engineering, University of Porto, Portugal

References

[1] Mautz, R., 2012. Indoor Positioning Technologies. Ph.D. thesis, ETH Zürich.

[2] Wan, E., Van Der Merwe, R., 2000. The unscented Kalman filter for nonlinear estimation. In: IEEE Adaptive Systems for Signal Processing, Communications, and Control Symposium.

[3] Thrun, S., 2002. Probabilistic robotics. Communications of the ACM.

[4] Lingemann, K., Nüchter, A., Hertzberg, J., Surmann, H., 2005. High-speed laser localization for mobile robots. In: Journal of Robotics and Autonomous Systems.

[5] Pomerleau, F., 2013. Methodology and Tools for ICP-like Algorithms. Ph.D. thesis, ETH Zürich.

[6] Magnusson, M., 2009. The three-dimensional normal-distributions transform - an efficient representation for registration, surface analysis, and loop detection. Ph.D. thesis, Örebro University.

[7] Kitt, B., Geiger, A., Lategahn, H., 2010. Visual odometry based on stereo image sequences with RANSAC-based outlier rejection scheme. In: IEEE International Conference on Intelligent Vehicles Symposium.

[8] Whelan, T., Johannsson, H., Kaess, M., Leonard, J.J., McDonald, J., 2013. Robust real-time visual odometry for dense RGB-D mapping. In: IEEE International Conference on Robotics and Automation.

[9] Costa, C., 2015. Robot Self-localization in dynamic environments. M.Eng thesis, FE-UP.

[10] Costa, C., Sobreira, H., Sousa, A., Veiga, G., 2015. Robust and accurate localization system for mobile manipulators in cluttered environments. IEEE International Conference on Industrial Technology.

[11] Quigley, M., Conley, K., Gerkey, B., Faust, J., Foote, T., Leibs, J., Berger, E. Wheeler, R., 2009. ROS: An open-source Robot Operating System. In: IEEE International Conference on Robotics and Automation.

[12] Rusu, R., Cousins, S., 2011. 3D is here: Point Cloud Library (PCL). In: IEEE International Conference on Robotics and Automation.

[13] Hornung, A., Wurm, K. M., Bennewitz, M., Stachniss, C., Burgard, W., 2013. OctoMap: An efficient probabilistic 3D mapping framework based on octrees. Journal of Autonomous Robots.

[14] Lowe, D., 2004. Distinctive image features from scale-invariant keypoints. Journal of Computer Vision.

[15] Zhong, Y., 2009. Intrinsic shape signatures: A shape descriptor for 3d object recognition. In: IEEE International Conference on Computer Vision.

[16] Rusu, R., Blodow, N., Marton, Z., Beetz, M., 2008. Aligning point cloud views using persistent feature histograms. In: IEEE/RSJ International Conference on Intelligent Robots and Systems.

[17] Rusu, R., Blodow, N., Beetz, M., 2009. Fast point feature histograms (FPFH) for 3d registration. In: IEEE International Conference on Robotics and Automation.

[18] Tombari, F., Salti, S., Di Stefano, L., 2011. A combined texture shape descriptor for enhanced 3d feature matching. In: IEEE International Conference on Image Processing.

[19] Frome, A., Huber, D., Kolluri, R., Bülow, T., Malik, J., 2004. Recognizing objects in range data using regional point descriptors. In: European Conference on Computer Vision.

[20] Tombari, F., Salti, S., Di Stefano, L., 2010. Unique shape context for 3d data description. In: Proceedings of the ACM Workshop on 3D Object Retrieval.

[21] Wohlkinger, W., Vincze, M., 2011. Ensemble of shape functions for 3d object classification. In: IEEE International Conference on Robotics and Biomimetics.

[22] Besl, P., McKay, N., 1992. A method for registration of 3-d shapes. In: IEEE Transactions on Pattern Analysis and Machine Intelligence.

[23] Segal, A., Haehnel, D., Thrun, S., 2009. Generalized-ICP. In: Proceedings of Robotics: Science and Systems.

[24] Sturm, J., Engelhard, N., Endres, F., Burgard, W., Cremers, D., 2012. A benchmark for the evaluation of RGB-D SLAM systems. In: IEEE/RSJ International Conference on Intelligent Robots and Systems.

[25] Pomerleau, F., Magnenat, S., Colas, F., Liu, M., Siegwart, R., 2011. Tracking a depth camera: Parameter exploration for fast ICP. In: IEEE/RSJ International Conference on Intelligent Robots and Systems.

Chapter 3

Rapid Prototyping of Embedded Video Processing Systems in FPGA Devices

Andrej Trost and Andrej Žemva

Additional information is available at the end of the chapter

http://dx.doi.org/10.5772/61136

Abstract

Design of video processing circuits requires a variety of tools and knowledge, and it is difficult to find the right combination of tools for an efficient design process, specifically when considering open tools for evaluation or educational purpose. This chapter presents an overview of video processing requirements, programmable devices used for embedded video processing and the components of a video processing chain. We propose a novel design flow for generating customizable intellectual property (IP) cores used in streaming video processing applications. This design flow is based on domain-specific modules in Python language. Examples of generated cores are presented.

Keywords: Video processing, prototyping, FPGA, Python, IP core

1. Introduction

Embedded sensor data processing is an important concept of future ubiquitous computing technology. Processing of data from video camera requires either a powerful and energy-consuming general purpose processor or an application-specific integrated circuit (ASIC), which is better suited for embedded applications. Field-programmable gate array (FPGA) technology enables rapid development and hardware prototyping of video processing ASICs. The FPGA devices are commonly found in smart camera architectures [1].

The technology is available, but it is not widely adopted due to the complex design process and the cost of specialized tools. A digital system design flow starts with hierarchy of circuit components containing register transfer level (RTL) description of the digital circuit in one of the hardware description languages VHDL or Verilog. The RTL circuit model is used for logic synthesis as well as simulation of the design. The synthesized design runs through implementation phase (mapping to the FPGA structures, place, and route, optimization), finally producing bitstream that is downloaded to the device. Compiling of the FPGA design and circuit debugging can be very time-consuming processes and large amount of logic cells and flip-flops in modern FPGAs call for an efficient design flow.

There are several solutions to speed up the circuit component development compared to the traditional RTL approach. Domain-specific environments, for example LabVIEW or System Generator for Digital Signal Processing in Matlab/Simulink, can be used to compose signal processing algorithm and produce RTL code. The other approaches are raising the abstraction level by using high-level software languages (C, SystemC, HandelC) and performing high-level circuit synthesis from software description [2]. The high-level tools are either in the research domain or in the commercial tools (for example, Vivado High Level Synthesis from Xilinx) and are targeting a variety of applications. There is still a lot of work for a circuit designer to adapt the synthesized IP components to the specific stream video processing architecture and application [3]. We also consider that open design environment is important for educational purpose and for wide adoption of the programmable technology [4].

In this chapter, we propose to use a high-level programming language Python to produce generators of video processing components. The Python language is efficient for algorithm development and visualization and provides modules for description of concurrent digital systems with automatic conversion to RTL. The Python scripts can be used to produce generic image processing intellectual property (IP) components that are customizable beyond the scope of the RTL languages. We introduce the design methodology and present the preliminary results.

In the next section, we present data processing and storage requirements in embedded video systems and examples of video processing chain. We show that even low end programmable devices from two major FPGA vendors, Xilinx and Altera, have enough resources and speed for real-time implementation of video processing chain.

In further sections, we give an overview of hardware description packages in Python and our IP generator module. The IP generator classes are used to automate several tasks and data transformations in the streaming hardware component development process. We show the usage of IP generator module on design examples.

2. Embedded video processing

Embedded video processing devices are part of smart cameras used in computer vision systems. Table 1 summarizes data processing and storage requirements for various video sources.

The minimal requirements for digital video processing systems are defined by standard definition PAL or NTSC video signals. Digital video stream from standard definition camera in ITU-R BT.656 format [5] is a sequence of 8-bit or 10-bit data words transmitted at 27 MB/s.

http://dx.doi.org/10.5772/61136

The video frame resolution for PAL camera is 720 by 576 pixels and one image frame occupies 405 kB of memory.

Video signal parameter	**PAL TV**	**WVGA**	**HD camera**
resolution [H x V]	720 x 576	752 x 480	2048 x 2048
refresh [fps]	25	63	178
data rate [MB/s]	27	27	760
frame memory [MB]	0.4	0.3	4

Table 1. Parameters of video signals

The next two columns present requirements for two commercial computer vision cameras: high-speed, low–resolution, and high-definition (HD) camera from Optomotive. The first has Wide Video Graphics Array (WVGA) sensor that has equal data rate and similar resolution to PAL camera but substantially higher refresh rate. The digitized PAL video stream contains long synchronization sequences due to the legacy TV transmission signal timing. On the other hand, the computer vision camera consumes most of the 27 MB/s bandwidth for data transmission and thus achieves refresh rates up to 63 frames per second (fps).

Characteristics of the HD camera from Table 1 show substantially higher resolution and roughly 10 times more memory storage per image frame. The sensor peak data rate of 760 MB/s enables image refresh rates of up to 178 fps.

The data rate of a digital video stream and the required amount of operations per image pixel are beyond the capacity of an embedded microprocessor. The video processing operations should be implemented in application-specific hardware. The blocks of the video processing chain operate concurrently in hardware and transformation operations inside each block can be parallelized.

2.1. Video processing chain

Video and image processing algorithms contain a sequence of data transformation operations. When the same sequence is applied to all image data, we can describe the algorithm as video processing chain. The processing chain clearly describes dataflow and can be used as template structure for hardware implementation. Structure of the data and the transformation operations depend on the video processing application [6].

Two examples of video processing chain are presented in Figure 1: (a) edge detection chain and (b) feature extraction chain for human detection. The edge detection transformations produce images that emphasize edges and transitions. The presented processing chain begins with contrast enhancement. The output pixel values are computed from the input pixels using contract stretching equation or lookup table. Next operations are gradient filters that produce local horizontal (Gx) and vertical (Gy) gradient data. The gradient data stream is aligned, combined in a single stream, and sent to gradient magnitude computation block.

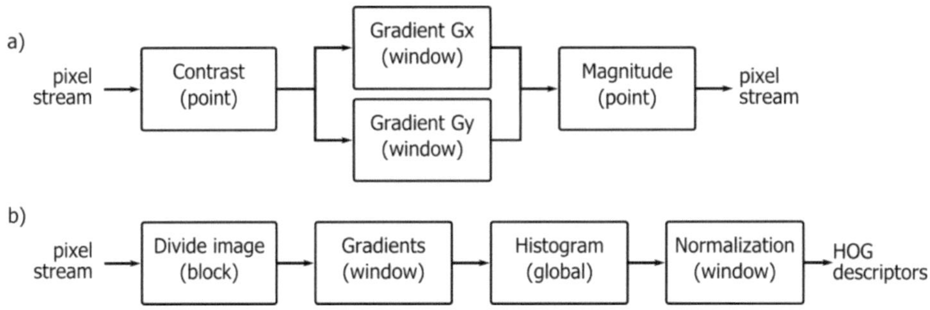

Figure 1. Examples of video processing chains

The feature extraction chain in Figure 1b transforms image pixel data to a set of descriptors used for classification and recognition in computer vision systems. The presented processing chain is based on histogram of oriented gradients (HOG) [7]. The image is first divided into blocks and gradients of the block pixels are computed. Histograms of gradient magnitude for spatial orientations are calculated next and block normalization is performed. The resulting data are image descriptors that can be used in the classification of algorithm to detect human locations.

We divide the video stream processing operations into four categories: point operations, sliding window operations, image block operations, and global operations:

- Point operations take one image pixel at a time and produce output values based on the current input pixel value. Examples of point operations are contract enhancement, image binarization (thresholding), and color conversion. These operations are relatively straightforward for hardware implementation using pipelined arithmetic operations or lookup table approach.
- Sliding window operations or local operations use a local neighborhood of pixels to produce the output. Examples of the sliding window operations are convolution-based operations for image filtering or calculating image gradients. The hardware implementation requires first-in first-out (FIFO) buffers to generate the neighborhood of pixels and pipelined arithmetic operations.
- The image block operations first divide the image frame into smaller blocks and then apply operation to the whole block of pixels. For example, HOG feature extraction divides the image into 64 × 128 pixel blocks. The hardware implementation of block operations on the video stream requires image buffers and blocks memory control logic. The implementation can exploit block-level parallelism in order to achieve the required throughput.
- Global operations require double buffering of the whole frame or block data, since there is no defined locality and any input pixel can be used to calculate the output. Example of global

operations is histogram calculation. These operations are most difficult to implement in the custom hardware at a reasonable cost and are typically implemented as a combination of software control and hardware accelerator.

Considering implementation of the video processing chain in hardware, we begin with initial digital system partitioning according to the dataflow and video transformation operations. The operations present hierarchical circuit blocks that can be reused for different applications.

3. Video processing in FPGA devices

The FPGA technology is used as a programmable alternative to ASICs. The FPGA devices can exploit high degree of data processing parallelism that is necessary for real-time video processing. The programmability of FPGA devices has many benefits in video processing applications due to the constant evolution of new algorithms and standards. This technology is well suited for smart cameras, where the image sampling and application-specific preprocessing are performed before data transmission to the host [8].

A drawback of FPGA devices is relatively high cost compared to massive produced ASICs and relatively complicated design flow compared to microprocessors. The programmable technology grows from simple logic replacement devices introduced in the late 1980s to powerful contemporary generic computation devices with plenty of interfaces, memory, and data processing resources. Even the smallest and low-cost devices today include static memory blocks and hardware support for fast arithmetic operations. In order to leverage usage of FPGA in embedded image processing applications, we consider low end devices and propose a novel design flow.

Table 2 presents characteristics of low end FPGA devices from two major manufacturers: Xilinx and Altera. A range of available resources in terms of logic cells, data flip-flops, embedded memory blocks, and hardware multiplier blocks are presented for FPGA families Cyclone IV and Max 10 from Altera and Spartan-6 and Artix from Xilinx.

FPGA family resource	**Altera Cyclone IV**	**Altera Max 10**	**Xilinx Spartan 6**	**Xilinx Artix**
logic cells [k]	6 – 149	2 – 50	3.8 – 147	16 – 215
flip-flops [k]	6 – 149	2 – 50	4.8 – 184	20 – 269
memory [kB]	33 – 486	13 – 204	27 – 603	112 – 1642
multipliers	15 – 266	16 – 144	8 – 180	45 – 740

Table 2. Low end FPGA devices from Altera and Xilinx

FPGAs are best suited for signal processing algorithms based on arbitrary precision integer or fixed point operations. The low end families have the smallest amount of data processing resources operating at a relatively low clock frequency (typically tens or few hundreds of MHz) due to programmability overhead. But even the smallest FPGA devices from the current families have enough resources and speed to implement real-time video image acquisition and some image processing functionality. Devices with more resources benefit from more parallelism that is important for HD or high frame rate video processing.

Emerging type of programmable integrated circuits are systems-on-chip (SoC), a combination of application-grade microprocessor and FPGA fabric. Both Altera and Xilinx provide SoC devices, Cyclone V and Zynq, which include dual core ARM Cortex-A9 processor. The smaller devices compete in terms of cost with the separate FPGA and microprocessor or microcontroller solution and benefit in tight coupling between processor and FPGA. These devices are suitable for embedded image processing applications that are partially implemented in hardware (HW) and in software (SW) and transfer data through microprocessor peripheral interfaces.

The design tools for programmable devices support component-based hierarchical design in order to manage development and verification of complex digital systems. The reusable blocks, called IP components, can be obtained from library of IP cores (Xilinx) or Megafunctions (Altera) or described using hardware description languages. Basic operations and components using specific FPGA structures are available for free, but a lot of IP cores can be obtained only after purchase.

While using IP cores for digital system shortens the design cycle, there is still a huge gap between the algorithm development and the circuit development. In the video processing algorithm development process, we consider new combinations of operations and different data partitioning to get an optimum processing result. New operators are probably not available as IP cores and need to be designed from scratch. The algorithm developer can also consider the cost of the hardware implementation of the operations in the design process. The algorithms are developed with tools and environments that offer strong mathematical and visualization support in order to get quick proof of concept. The tools are either commercial Matlab or LabVIEW or based on computer languages C/C++ (OpenCV) or Python.

3.1. Verification on development boards

For video processing hardware development, we can use either a computer vision smart camera with programmable device or a programmable video development board. In order to introduce design of embedded video processing in university laboratory practice, we developed video interface modules that can be used with low-cost commercial FPGA development boards. This solution is more affordable and the interface modules can be reused when the FPGA vendors offer new development boards based on new families of FPGA devices.

The video input interface module contains video decoder TVP5150A connected to PAL camera module, as presented in Figure 2. A small FPGA is used for the decoder setup, basic data preprocessing, and output video stream configuration. The video decoder TVP5150A converts analog video to the video stream in TU-R BT656 format. The FPGA device Xilinx XC3S50A performs stream decomposition and color space conversion and generates data stream. The

http://dx.doi.org/10.5772/61136

board has a 40-pin parallel video stream connector and a 12-pin serial PMOD for connection to low-cost Xilinx Digilent boards.

Figure 2. Photo of video interface board

The Digilent FPGA development boards for Xilinx FPGAs contain a computer graphics VGA or HDMI connector that can be used as a visual output for the image processing application. If there is no video output on the board, we can always add a simple extension module with the computer graphics or standard video output.

4. Hardware description in Python

The open-source Python community offers several packages covering digital circuits design. We tested three packages that use Python as a hardware description language [9].

Chips [10] is an HDL Python library that provides a language for designing hardware devices. Chips package introduces a stand-alone synthesizable language built on top of the Python allowing designers to work at a higher level of abstraction. The language provides methods for concurrent elements synchronization and data stream processing. The tool generates synthesizable RTL code using state machines or optimized soft-core processor automatically. The described device can be natively simulated in Python.

Migen [11] is another Python-based tool for building complex digital hardware. The toolbox introduces inside Python a new language FHDL for describing fully synchronous circuits. The language addresses limitations of standard hardware description languages: support for composite types and procedurally generated logic. The FHDL circuit description is on a higher abstraction level compared to the RTL languages and can be automatically converted to synthesizable Verilog. The simulation is supported through conversion and linked to external tools.

MyHDL [12] is an open-source package for using Python as a hardware description and verification language. The code can be converted to VHDL or Verilog automatically. The introduced language MyHDL does not specifically target synchronous or stream processing circuits and is intended for the description of synchronous or asynchronous logic blocks. The converted code is readable, since it retains all the signal and component names and even block comments. The MyHDL supports native Python simulations and unit tests and can produce outputs in value change dump (VCD) format for graphical inspection. The package MyHDL provides basic RTL modeling concepts:

- Hardware-oriented data types: 1-bit bool and arbitrary length vectors intbv, modbv
- Support for synthesizable subset of arithmetic and logic expressions
- Combinational functions containing concurrent signal assignments
- Synchronous sequential functions with clock edge and optional reset
- Finite state machine abstraction
- Structural modeling.

Table 3 summarizes the basic features of the Python HDL tools. We found the tool MyHDL as the best choice for the development of reusable components due to nice modeling, conversion, simulation, and continuous support by the open-source community.

Package	Circuit model	Verification	Output	Current version
Chips	stream model, custom syntax	Python testbench	VHDL	0.1.2 (2011)
Migen	fragment model, custom syntax	External tools	Verilog	x (2012)
MyHDL	event driven, Python syntax	Python testbench and VCD	VHDL and Verilog	0.9 (2013)

Table 3. Python HDL Tool packages

4.1. MyHDL video graphics example

We will first present an example of a VGA graphics controller designed for ZynqSoC device. The controller is used to produce computer-generated video stream for hardware verification of the video processing chain on the FPGA development board. The controller presented in Figure 3 consists of device library components: Zynq PS, AXI interconnect, BRAM controller, Block memory, and custom components designed in MyHDL. The video stream generator components are VGA synchronization (VGAsync), video direct memory access (DMA), coordinate rotation, character memory, and color transformation.

We present the MyHDL code and hardware modeling constructs on a simplified example of the VGA synchronization generator:

http://dx.doi.org/10.5772/61136

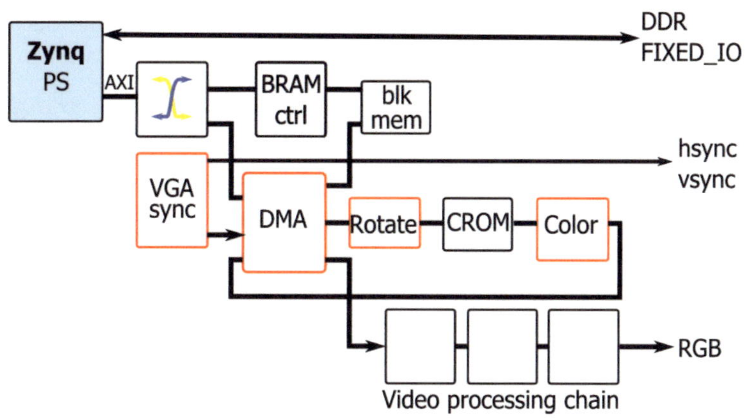

Figure 3. Video processing rapid prototyping components in ZynqSoC

```
from myhdl import *              # libraries and initialization code
HP=1040; VP=666
def VGAsync(clk, hsync, vsync):          # top level function and ports
        h = Signal(intbv(0)[11:])         # internal signals
        v = Signal(intbv(0)[10:])
        @always(clk.posedge)             # sequential function
def timing():
if h < HP-1:
                     h.next = h + 1
else:
                     h.next = 0
if v < VP-1:
                     v.next = v + 1
else:
                     v.next = 0
@always_comb                        # combinational function
def synchro():
hsync.next = 1 if h>=856 and h<976 else 0
vsync.next = 1 if v>=637 and v<643 else 0
return timing, synchro
clk = Signal(bool(0))              # port signal declaration
hsync = Signal(bool(0))
vsync = Signal(bool(0))
if __name__ == '__main__':      # conversion to VHDL
         toVHDL(VGAsync, clk, hsync, vsync)
```

The MyHDL hardware model contains declaration of signal objects (Signal) and functions describing combinational or synchronous sequential logic using Python decorators (@always). Integer values can be used for single bit signals (bool) as well as for bit vectors (intbv) which greatly simplifies the code compared to strict VHDL typing rules. The MyHDL hardware description requires less code compared to automatically generated or even handwritten VHDL or Verilog description. We can write a test bench using MyHDL objects and verify the operation of the design in Python. The verification is performed by printing the consecutive signal values or dumping all the data to a timing waveform file. Python scripts can be written for automatic verification and unit testing.

The RTL circuit description requires careful design of the circuit parts with specified clock cycle behavior, such as interfaces. Figure 4 presents simplified algorithmic state diagram of the region movement DMA component. The state machine leaves initial IDLE state when a pixel read request (rdreq) or write request (wrreq) is received. The read request is asserted for each image line, and during video blanking period the controller returns to IDLE state. When the controller receives write request and is not reading the memory, it starts data movement by first getting the data from one bus (state GET) and then writing the data to frame memory (state PUT). The internal counter dxy is used to repeat these cycles and move 64 data pixels. The data movement cycle can be interrupted by read request and continued when the rdreq is de-asserted.

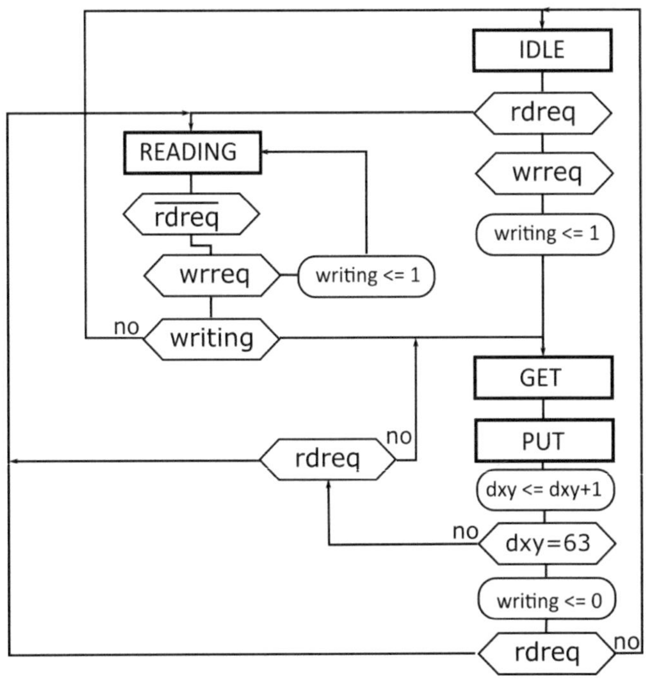

Figure 4. Video Direct Memory Access (DMA) controller state diagram

The RTL implementation of the DMA requires the designer to specify all the control and status signals of the data busses involved in the communication. For example, the MyHDL description of IDLE and GET is stated as follows:

```
if st==tst.IDLE:                  # bus1 write request
if stb_i&we_i:
        writing.next=1
        stb2_o.next=1; we2_o.next=0;      # start reading cycle on bus2
```

```
        st.next=tst.GET
...
elif st==tst.GET:
        stb2_o.next=1; we2_o.next=0      # continue reading on bus2
if ack2_i:                        # if read acknowledge
        stb2_o.next=0                    # stop reading bus2 and
        stb3_o.next=1; we3_o.next=1      # start writing cycle on bus3
        dat3_o.next=data;
        adr3_o.next = concat(adrH, adrL)
        st.next=tst.PUT
```

The presented circuit description is time consuming and error-prone process and can be avoided by using higher level of abstraction for the specific domain.

5. IP component generator in Python

An IP component generator written in Python language can be used to raise the MyHDL hardware description level of abstraction and target video stream processing IP components. The proposed design flow is presented in Figure 5. We created a collection of Python classes in IPgen module used for object-oriented hardware description. The IP designer creates the IP generator script file composed of IPgen objects and methods. When the script is executed, transformations generate IP description file with RTL circuit description in MyHDL. This Python file can be used for functional IP verification with test bench and automatic conversion to HDL code.

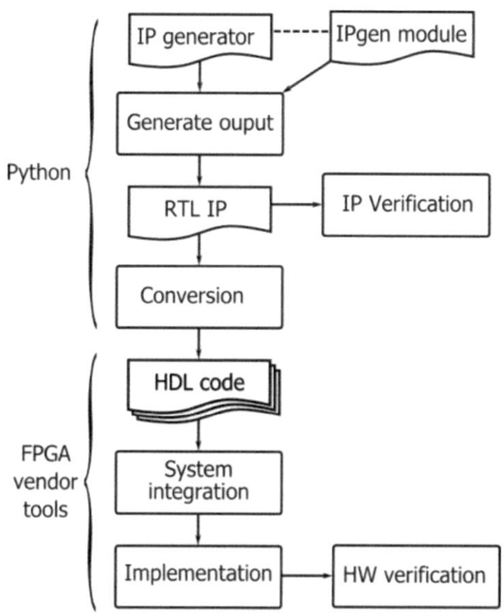

Figure 5. Python based IP generator design flow

The resulting HDL code presents RTL description of one IP in the video processing chain. The FPGA vendor tools are then used to package IP, integrate IP in the video processing hardware system, and perform circuit implementation. The implemented circuit is downloaded to FPGA for hardware verification on video development board.

5.1. IP generator module

First, we present semantics of object-oriented hardware description in our IP generator module. We define the IP component as a tuple where *IPname* stores top level function name, *IPinit* defines initialization code, and *IPinterface* defines interface type. The IP component contains a set of functions and a set of signals.

Statements are a collection of MyHDL statement code and identifiers. The identifiers are used in code transformation process. The set of signals is defined with name, type, and scope for each signal. The signal type is one of the MyHDL provided types. The scope defines the position of the signal declaration:

- External signals are part of initialization code.
- Port signals are on port description of the top level function.
- Internal signals are declared inside top level function.

A function transformation converts function to a set of generated functions and signals:

$$generate(\mathcal{F}) \rightarrow \{(name_0, \mathcal{F}_{g0}, \mathcal{S}_{g0}), \ldots, (name_k, \mathcal{F}_{gk}, \mathcal{S}_{gk})\}$$

The generated functions are objects of the basic combinational or sequential MyHDL functions and can be converted to MyHDL code using *getcode()* method. The generated signal objects have *declare()* method that outputs appropriate signal declarations. The algorithm for producing MyHDL IP description is presented in pseudocode:

```
init_code = IPinit                    # get initialization code
(iname, if, sig) = IPif.generate()    # generate interface logic
F.append(if)                           # and append to functions list

for fun in F:                         # for each function generate
     (fname, ff, fsig) = fun.generate()    # list of function names
     lname += fname;                       # list of function names
     sig += fsig;                     # list of signals
     code += ff.getcode()             # and function code

for sig in S:                         # for each signal generate
if sig.stype==port:
     port_code += sig.name;                # list of port names and
     psig_decl += sig.declare()            # port declarations
else:
     sig_decl += sig.declare()             # internal signal declarations
```

http://dx.doi.org/10.5772/61136

```
print init_code                      # print out initialization code
print "def" + name + port_code        # top level function and ports
print sig_decl                       # signal declarations
print code                           # function code
print "return" + lname               # top level return
print psig_decl                      # port signal declarations
print conversion function            # and conversion to VHDL/Verilog
```

The structure of our stream video processing domain module is presented in UML diagram in Figure 6. Python classes are provided for description of the IP component that is an object of type IPgen. The object IPgen is a collection of signal and function objects defining the IP hardware structure. A name property is used for MyHDL top level function as well as for output file name for MyHDL circuit description. The interface property provides various automatically generated IP component interfaces for streaming video components.

The IPgen signals are external or internal signals described in integer (si) or binary (sb) signal objects. The classes provide declaration method that is used to generate MyHDL signal declarations. The IPgen class *function()* is basic class for description of the IP behavior. The *code()* method is used to add MyHDL statements into statement code list (C[]). The function code is translated into a set of MyHDL combinational and sequential functions using *getcode()* method. All classes derived from the *function()* should provide the *generate()* method that returns a triple containing list of MyHDL function names, list of basic functions objects, and list of signals.

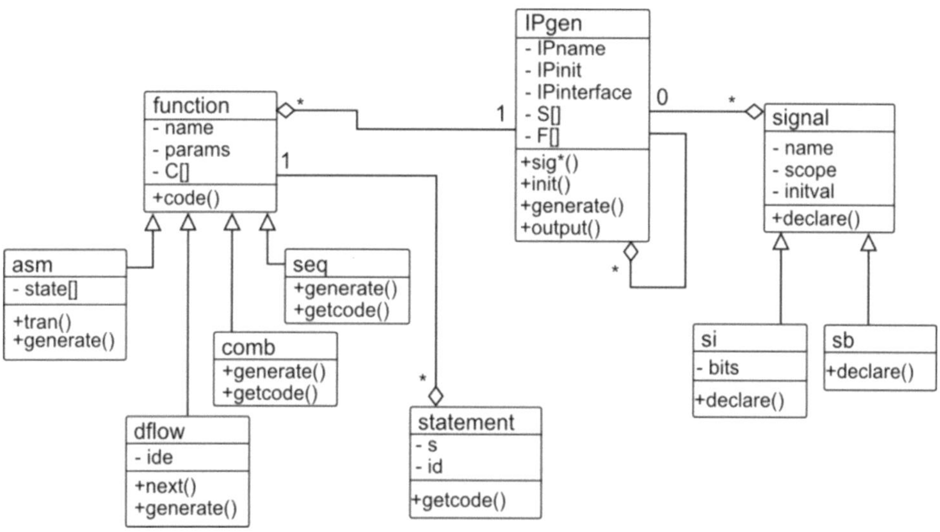

Figure 6. UML diagram of Python IP generator module

The presented generic MyHDL IP generator module can be extended by domain-specific classes describing interface and functions objects. We developed extensions for streaming video processing applications that are presented in the design examples.

5.2. IP generator design examples

We present IP description of square root operation on integer data that is commonly found in image processing transformations. The code is based on modified Dijkstra's square root algorithm [13]. The square root in plain Python code is calculated in a loop:

```
def sqr(number, bits):
     mask = 1 << (bits-2)
     r = 0
     I = 1
     rem = number
while (mask>0):

if ((r+mask)<=rem):
           rem -= r+mask
           r += 2*mask
       r >>= 1
       mask>>= 2

return r
```

The loop has a constant number of iterations that can be calculated from the input number bit size, for example, 8-bit numbers require 4 loop iterations. The square root of an 8-bit input data can be calculated as a sequence of operations presented in Figure 7a. The combinational functions of the corresponding MyHDL circuit description are given in Figure 7b.

The circuit description can be generated with a script using IP generator classes. We first define IPgen object and declare arrays of vector signals in dataflow diagram:

```
bits = 8
w = (bits-2)/2 + 1
ip = IPgen("sqr", IFpoint) #set the IP name and interface type
ip.siga('r', w, bits)
ip.siga('m', w, bits)
ip.siga('rem', w, bits)
```

A member function siga() is used for declaration of multiple signal instances; for example, a function call siga('r', 4, 8) produces the following MyHDL declaration:

```
r0, r1, r2, r3 = (Signal(intbv(0)[8:]) for i in range(0,4))
```

The initialization logic is described as:

```
log = comb("init_logic")
log.code("rem0.next = data")
log.code("r0.next = 0")
log.code("m0.next = 1 << {0}".format(bits-2))
```

The code for dataflow object is generated in a for-loop:

```
p = dflow("p",DFcomb)
for i inrange (0, w):
```

http://dx.doi.org/10.5772/61136

a)

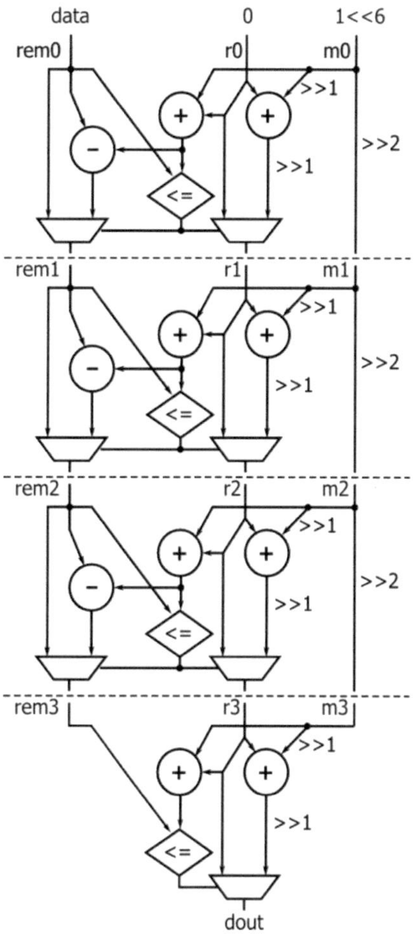

b)

```
@always_comb
def init_logic():
    rem0.next = data
    r0.next = 0
    m0.next = 1 << 6

@always_comb
def p_logic():
    if (r0+m0) <= rem0:
        rem1.next = rem0-(r0+m0)
        r1.next = (r0 + 2*m0 ) >> 1
    else:
        rem1.next = rem0
        r1.next = r0 >> 1
    m1.next = m0 >> 2
    if (r1+m1) <= rem1:
        rem2.next = rem1-(r1+m1)
        r2.next = (r1 + 2*m1 ) >> 1
    else:
        rem2.next = rem1
        r2.next = r1 >> 1
    m2.next = m1 >> 2
    if (r2+m2) <= rem2:
        rem3.next = rem2-(r2+m2)
        r3.next = (r2 + 2*m2 ) >> 1
    else:
        rem3.next = rem2
        r3.next = r2 >> 1
    m3.next = m2 >> 2
    if (r3+m3) <= rem3:
        dout.next = (r3 + 2*m3 ) >> 1
    else:
        dout.next = r3 >> 1
```

Figure 7. a) 8-bit integer square root dataflow, b) MyHDL combinational square root description

```
        p.code("if (r{0}+m{0}) <= rem{0}:".format(i))
    if (i == w-1):
            p.code(" dout.next = (r{0} + 2*m{0}) >> 1".format(i))
    else:
            p.code(" rem{1}.next = rem{0}-(r{0}+m{0})".format(i, i+1))
            p.code(" r{1}.next = (r{0} + 2*m{0}) >> 1".format(i, i+1))
            p.code("else:")
    if (i == w-1):
            p.code(" dout.next = r{0} >> 1".format(i))
    else:
            p.code(" rem{1}.next = rem{0}".format(i, i+1))
            p.code(" r{1}.next = r{0} >> 1".format(i, i+1))
            p.code("m{1}.next = m{0} >> 2".format(i, i+1))
            p.next()
```

In the dataflow loop description, we call a method next() to mark computation stages denoted with a dashed line in Figure 7a. The stages are used in the automatic generator of RTL combinational or sequential dataflow descriptions. The dataflow object accepts parameter that defines the dataflow synthesis mode with the following values:

- DFcomb for combinational logic implementation (for example in Figure 7b)
- DFreg for combinational description with a register in the last stage
- DFpipe for sequential description with registers in all stages.

The RTL IP component description in MyHDL is generated by

```
ip.output(log, p)
```

By changing the parameter in the dataflow object, the user can quickly produce different versions of the circuit and choose the version that satisfies timing and area constraints. The output description includes generated interface control logic for the video stream IP.

The IP core interfaces in digital systems are either generic function-specific interfaces or standard bus interfaces. The standard bus interfaces are based on proprietary processor bus architectures (Avalon, AXI, PLB) or open architectures for digital systems (Wishbone). We consider using a simplified version of WISHBONE [14] interface specifically tailored for video data stream transmission. The video stream interface should provide variable size pixel data bus, clock, and data strobe signal used to mark active data transmission cycle. The proposed video bus timing waveform for the interface type IFpoint is presented in Figure 8.

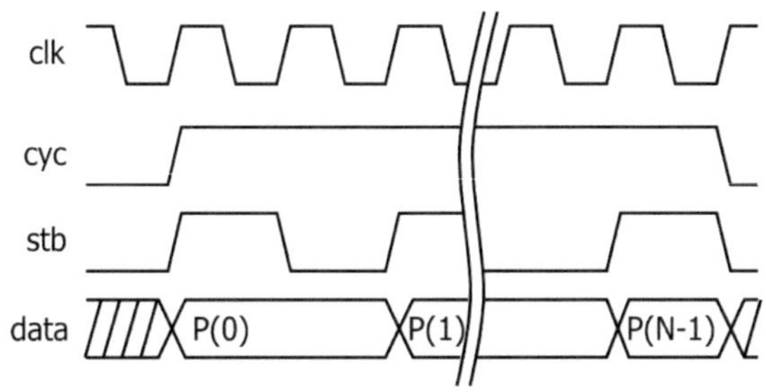

Figure 8. Video data stream bus timing waveform

The interface signals and control logic are automatically generated. The control logic is used to count pipeline cycles, produce output strobe for valid output data, and provide additional cycles to finish the pipeline computation of the data from the input burst. The generated interface code for four-stage pipeline of 8-bit square root circuit is

```
        # Interface internal signal declarations
        pipe = Signal(bool(0))          # enable additional cycles
        cycEnd = Signal(bool(0))         # end output cycle
        cycnt = Signal(intbv(0)[8:])     # cycle counter
...
```

http://dx.doi.org/10.5772/61136

```
          # Interface control logic
          stbo.next = 0; pipe.next = 0; cycEnd.next = 0
if stb and cyc:                       # count pipeline cycles
if cycnt == 3:
               stbo.next = 1
               cyco.next = 1
else:
               cycnt.next = cycnt + 1
elif (not stb) and (not cyc): # additional pipe cycles
if cycnt> 0:
               cycnt.next = cycnt - 1
               pipe.next = 1
if pipe:
          stbo.next = 1
if cycnt == 0:
               cycEnd.next = 1
if cycEnd == 1:
          cyco.next = 0
```

The initial Python description of the square root calculation is 35 lines of code and is automatically transformed to 72 lines of MyHDL and 126 lines of VHDL or Verilog code.

The next example is a sliding window operation for image filtering. The filtering is performed by a discrete convolution of the input data with filter coefficients C_0, C_1,...,C_n. Horizontal convolution equation in *Z* domain:

$$P_{out} = C_0 P + C_1 P z^{-1} + C_2 P z^{-2} + \cdots + C_n P z^{-n}$$

The convolution can be decomposed into multiply–add operators and delay elements (z^{-1}). The convolution core is generated with one combinational and one sequential function:

```
n = 3
c = [1, 2, 1]
log = comb("filt_log")
reg = seq("filt_reg")
reg.code("if en:");
for i inrange (0, n):
     log.code("r{0}.next = s{0} + {2}*data".format(i,i+1,c[i]))
if (i < n-1):
          reg.code(" s{1}.next = r{0}".format(i,i+1))
```

For implementation of the filter in stream processing IP component, we have to add interface logic and consider boundary conditions in computation of the output values. Pixels outside image boundary should have zero value in convolution computation. To insert this data into convolution circuit core, we add an algorithmic state machine (ASM) driving data multiplexer and input FIFO buffer. The dataflow structure of the stream processing filter is presented in Figure 9.

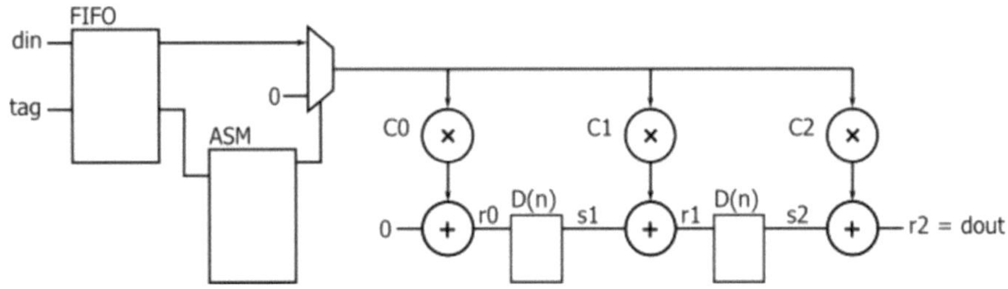

Figure 9. Dataflow diagram for convolution filter with 3 coefficients

The image boundary pixels are marked with additional tag signal on the input. When the first line pixel is received, the ASM switches input multiplexer to zero value and starts computation cycles. The FIFO is used to hold incoming pixels before they are used in convolution computation. Similar sequence is performed for the last line pixel. State diagram and the corresponding IP generator code of the ASM is presented in Figure 10.

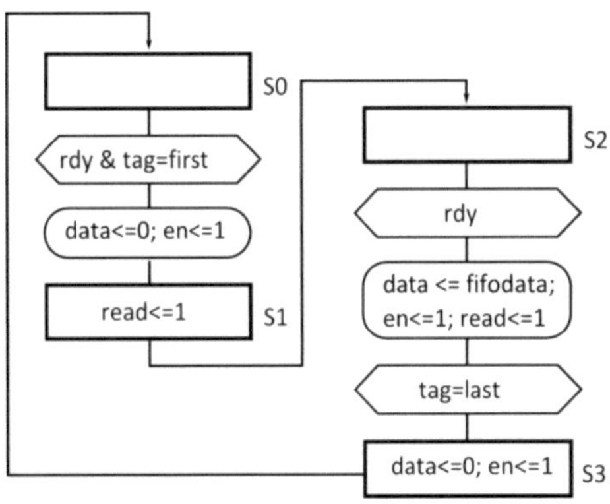


```
a=asm("st","S0")
a.default("en.next=0; read.next=0")

a.tran("S0", "", "rdy and tag==FIRST",
 "data.next=0; en.next=1", "S1")
a.tran("S1", "read.next=1", "", "", "S2")
a.tran("S2", "", "rdy", "data.next=fifodata;
 en.next=1; read.next=1", "")
a.tran("S2", "", "rdy and tag==LAST","","S3")
a.tran("S3", "data.next=0; en.next=1", "",
 "", "S0")
```

Figure 10. Algorithmic state machine for the convolution controller

The ASM states and state transitions are described with tran() method using five parameters: current state, state logic, transition condition, transition logic, and next state. For example, in state S0 the transition condition is FIFO ready signal (rdy) and tag that marks the first pixel in a line. If the condition is satisfied, the data signal for the convolution core is set to 0 and one convolution cycle is enabled (en). The default value of control signals en and FIFO read is 0 and the signals are activated only in specified state or transition logic. The ASM description is automatically converted to MyHDL sequential function and state machine signal declarations.

The FIFO buffer description and interface logic is generated by setting sliding window interface in IPgen object. The generated MyHDL code for convolution filter with three coefficients contains 115 lines and is transformed to 160 lines of VHDL or Verilog. Vertical convolution is computed by replacing the delay registers D(n) in the convolution core with circular buffers for one image line. The ASM logic should be changed to detect first and last image line and replace the boundary lines with zero values.

6. Conclusion

We presented unified embedded video processing circuit design flow in the Python language that speeds up the design cycle and automates error-prone domain conversion process. Python scripting language is a good choice for testing domain-specific algorithm and data transformations using open-source packages. We selected the package MyHDL for RTL hardware description and automatic conversion to readable VHDL or Verilog code. The advantage of MyHDL in comparison to traditional hardware description languages is the ability to use all the scripting power of the language for unit testing or simulation verification and to raise the level description abstraction in the same environment. The drawbacks of MyHDL code are limited support of some advanced HDL features from standard language packages and complex error reporting. We overcome this limitation by using MyHDL as intermediate language and use the proposed IPgen module for the streaming IP component hardware description.

The IPgen module has generic object structures used for generating MyHDL components with combinational and sequential functions and domain-specific streaming functions. Code transformations in the output MyHDL generation process are used to determine the produced circuit timing behavior and add video processing interfaces, components, and signals. We presented examples of square root calculation and convolution filter designs. For the future work, we will extend the IPgen classes to include more interface options and provide resource mapping transformations.

Author details

Andrej Trost* and Andrej Žemva

*Address all correspondence to: andrej.trost@fe.uni-lj.si

Faculty of Electrical Engineering, University of Ljubljana, Ljubljana, Slovenia

References

[1] A.W. Malik, B. Tho®rnberg, and P. Kumar. Comparison of Three Smart Camera Architectures for Real-Time Machine Vision System. *Int J Adv Robot Syst*. 2013;10(402): 1–12. DOI: 10.5772/57135

[2] J. Cong, B. Liu, S. Neuendorffer, J. Noguera, K. Vissers, and Z. Zhang. High-Level Synthesis for FPGAs: From Prototyping to Deployment. *IEEE Transactions on Computer-Aided Design of Integrated Circuits and Systems*. 2011;30(4):473–491.

[3] J. Serot, F. Berry, and S. Ahmed. Implementing Stream-Processing Applications on FPGAs : A DSL-Based Approach. In: *Field Programmable Logic and Applications, FPL*; 2011; Chania, Crete, Greece. pp. 130–137.

[4] A. Trost and A. Žemva. Teaching Design of Video Processing Circuits. *International Journal of Electrical Engineering Education*. 2012;49(2):170–178. DOI: 10.7227/IJEEE.49.2.7

[5] International Telecommunication Union. Recommendation ITU-R BT.656-5 [Internet]. 2007. Available from: http://www.itu.int/ [Accessed: February 2015].

[6] D.G. Bailey. *Design for Embedded Image Processing on FPGAs*. John Wiley & Sons (Asia) Pte Ltd, Singapore; 2011. 416 p

[7] N. Dalal and B. Triggs. Histograms of Oriented Gradients for Human Detection. In: *Histograms of Oriented Gradients for Human Detection*; 2005; IEEE Computer Society Washington. pp. 886–893. DOI: 10.1109/CVPR.2005.177

[8] J. Sérot, F. Berry, and C. Bourrasset. High-Level Dataflow Programming for Real-Time Image Processing on Smart Cameras. *Journal of Real-Time Image Processing*. 2014;:1–13. DOI: 10.1007/s11554-014-0462-6

[9] A. Trost. Design of a Graphical Controller with Reusable Components. In: B. Zajc and A. Trost, editors. *Proceedings of the 22nd international Electrotechnical and Computer Science Conference* ERK 2013; September 2013; Portorož, Slovenia. p. 31–34.

[10] J. Dawson. Chips—Hardware Design in Python [Internet]. 2011. Available from: http://dawsonjon.github.io/chips/ [Accessed: February 2015].

[11] S. Bourdeauducq. Migen Manual [Internet]. March 2014. Available from: http://m-labs.hk/migen.pdf [Accessed: February 2015].

[12] J.I. Villar,J. Juan, M.J. Bellido, J. Viejo, D. Guerrero, and J. Decaluwe. Python as a Hardware Description Language: A Case Study. In: *VII Southern Conference on Programmable Logic (SPL)*; 2011; Cordoba. pp. 117–122.

[13] M.T. Tommiska. Area-Efficient Implementation of a Fast Square Root Algorithm. In: *Proceedings of Third IEEE International Caracas Conference on Devices, Circuits and Systems*; 2000; Caracas. pp. S18/1–S18/4.

[14] OpenCores. WISHBONE System-on-Chip (SoC) Interconnection Architecture for Portable IP Cores [Internet]. 2002. Available from: http://cdn.opencores.org/downloads/wbspec_b3.pdf [Accessed: February 2015].

Chapter 4

An Analysis of Data Link Control Protocols

Pinaki Mitra

Additional information is available at the end of the chapter

http://dx.doi.org/10.5772/60962

Abstract

In this chapter we analyze the performance of data link protocols. We consider Stop and Wait ARQ and Sliding Window protocol. In Sliding window protocol, we consider Selective Reject ARQ and Go-Back-N ARQ. There are existing results that analyze the link utilization of these protocols in the presence of error. We have experimented in a somewhat more generalized framework. We have also analyzed and experimented with the number of duplicate error-free packets that will be transmitted.

Keywords: Stop ad Wait ARQ, Sliding Window Protocol, Selective Reject ARQ, Go-Back-N ARQ, Duplicate Error-Free Packet

1. Introduction

There are two types of Data Link Control Protocols:

1. Stop and Wait
2. Sliding Window

1.1. Stop and wait ARQ

In stop and wait flow control the sender sends the data frame whose transmission time is t_f and it reaches the receiver after time t_p, the propagation delay across the communication channel. The receiver sends the acknowledgment frame back to the sender whose transmission time is negligible and the propagation delay is t_p. This is illustrated in Figure 1. The sender sends the next data frame after time ($t_f + 2t_p$) if ACK is positive, otherwise it resends the earlier frame.

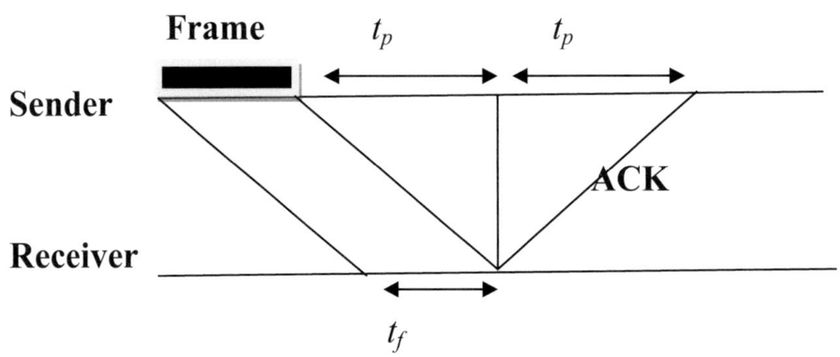

Figure 1. Illustration of Stop and Wait ARQ

The link utilization of this stop and wait protocol is given below:

$$U = \frac{t_f}{t_f + 2t_p}$$

1.2. Sliding window protocol

Here, the sender maintains a window of data frames of size W. In this protocol the sender continues sending packets before the receipt of the ACK (acknowledgment) frame. If the frame transmission time is t_f and the propagation delay is t_p, we can pack A data frames in the time span $t_f + 2t_p$ where

$$A = \frac{t_f + 2t_p}{t_f} = 1 + 2\frac{t_p}{t_f} = 1 + 2a \text{ where } a = \frac{t_p}{t_f}$$

If our window size $W \geq A$ then link utilization $U = 1$ otherwise $U = \frac{W}{A}$. The protocol is illustrated in Figure 2.

This analysis holds for error free transmission. Now if there are transmission errors then in both the protocols the packet has to be resent. For sliding window protocols there are two types of ARQ, namely:

1. Selective Reject ARQ

2. Go-Back-N ARQ

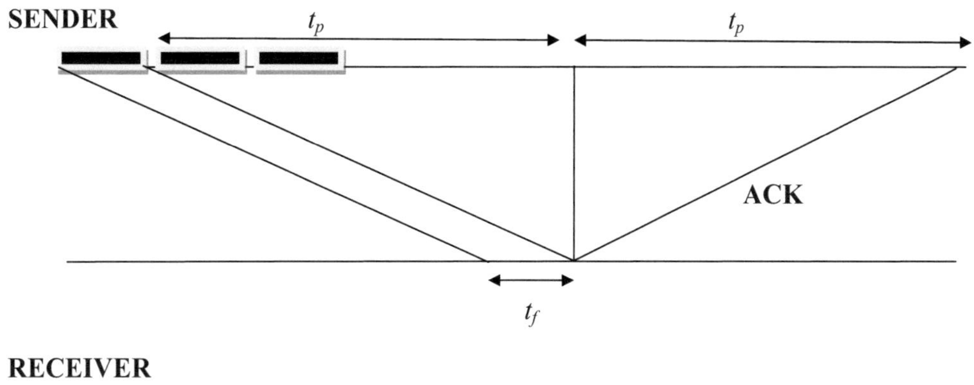

Figure 2. Illustration of Sliding Window Protocol

In Selective Reject ARQ, sliding windows of same size at both sending and receiving ends are maintained. The receiver sends a *RR-N* frame, that is, receive ready frame to the sender for the sending data frame *N*. If any data frame *N* is in error the receiver sends back *SREJ -N*, that is, selective reject signal to the sender. It continues to receive frames with sequence number ≥ *N* from the sender but it does not send any *RR* request for frame ≥ *N*.

In contrast, in Go-Back-N ARQ, sliding windows of unequal size at sending and receiving ends are maintained. In the sending end sliding window size is usually > 1 but in the receiving end sliding window size is usually = 1. Here also the receiver sends a *RR-N* frame, that is, receive ready frame to the sender for the sending data frame *N*. If any data frame *N* is in error, the receiver sends back *Go-Back-N* signal to the sender. It rejects all frames with sequence number ≥ *N* from the sender and it does not send any *RR* request for frame ≥ *N*.

For both these protocols, let N_r be the expected number of retransmissions. Then, the link utilization in the presence of error $U_{error} = \frac{U}{N_r}$. The analysis of N_r is usually carried out by assuming the ACK frame is error free. In this chapter, we try to analyze three situations:

1. The delivered packet is delayed
2. ACK frame is in error
3. ACK frame is delayed

In all the three cases the receiver will receive duplicate error-free packets. We try to analyze the estimate of duplicate packets received by the receiver.

2. Existing analysis of link utilization of data link control protocols

Consider *P* is the probability that the single frame is in error. Then link utilization for **Stop-and-Wait ARQ** is computed as follows:

$$N_r = \sum_{i=1}^{\infty}\left(iP^{i-1}(1-P)\right) = 1/(1-P)$$
$$U_{error} = U / N_r = (1-P)/A$$

Where

$$A = \frac{t_f + 2t_p}{t_f} = 1 + 2\frac{t_p}{t_f} = 1 + 2a \text{ where } a = \frac{t_p}{t_f}$$

For Selective Reject ARQ the link utilization is as follows:

$$U_{error} = U / N_r = \begin{cases} 1-P & W \geq A \\ \dfrac{W(1-P)}{A} & W < A \end{cases}$$

For Go-Back-N ARQ each error generates a requirement to retransmit K frames rather than a single frame.

$$N_r = \sum_{i=1}^{\infty}\left[1+(i-1)K\,P^{i-1}(1-P)\right] = 1 - K + K/(1-P) = (1-P+KP)/(1-P)$$

$$U_{error} = U / N_r = \begin{cases} \dfrac{1-P}{1+2aP} & W \geq A \\ \dfrac{W(1-P)}{A(1-P+WP)} & W < A \end{cases}$$

Here, we assumed $K = A$ for $W \geq A$ and $K = W$ for $W < A$.

A detailed discussion of the protocol and the above analysis can be found in authored books [1], [2], [3] and [4]. Reader can also refer to the published journal article [5].

3. Modified analysis of link utilization of data link control protocols

We will assume that the probability of a frame not in error is p_1. Also we will assume that the probability the frame is delivered within a time interval T is p_2. The sender will resend the packet if ACK does not arrive after time interval 2T. Let p_3 be the probability that the ACK frame is error free. Let p_4 be the probability that the ACK arrives within time 2T. The probability that the packet is received correctly in one transmission is $p_1p_2p_3p_4$. So the expected number of

http://dx.doi.org/10.5772/60962

retransmissions $N_r = 1/\ p_1p_2p_3p_4$. So in the existing analysis we only need to substitute $P = 1 - p_1p_2p_3p_4$ to get the link utilization.

To analyze the case of the receipt of duplicate error free packets we consider the following four events:

i. The delivered packet is free from error. The corresponding probability is p_1. Let us call this as event A_1.

ii. The packet doesn't reach within time T and the receiver asks to resend the packet. The corresponding probability is $(1-p_2)$. Let us call this as event A_2.

iii. The ACK frame is in error and requires retransmission. The corresponding probability is $(1-p_3)$. Let us call this as event A_3.

iv. The ACK frame doesn't arrive within time 2T and the sender resends the packet. The corresponding probability is $(1-p_4)$. Let us call this as event A_4.

The probability that there is duplicate frame transmitted will be:

$$\begin{aligned}
&\mathrm{P}[A_1 \cap (A_2 \cup A_3 \cup A_4)] = \mathrm{P}(A_1)\left[\mathrm{P}(A_2) + \mathrm{P}(A_3) + \right.\\
&\mathrm{P}(A_4) - \mathrm{P}(A_2 \cap A_3) - \mathrm{P}(A_3 \cap A_4) - \mathrm{P}(A_4 \cap A_2) - \mathrm{P}(A_2 \cap A_3 \cap A_4)\,]\\
&= p_1 \begin{bmatrix} (1-p_2) + (1-p_3) + (1-p_4) - (1-p_2)(1-p_3) - \\ (1-p_3)(1-p_4) - (1-p_4)(1-p_2) + (1-p_2)(1-p_3)(1-p_4) \end{bmatrix} = q
\end{aligned}$$

So the expected number of duplicate packets will be $N_r\, q$.

4. Experimental results and discussions

We have experimented with the number of retransmission and the number of duplicate error-free packets transmitted with the following C code:

```
#include <stdio.h>
#include <math.h>
#include <stdlib.h>
#include <time.h>
void main()
{
float p[4];
float S0, S1, S2;
int i,j;
float q, term, term_1, term_2;
int n_r, dup;
time_t t;
srand((unsigned) time(&t));
```

```
randomize();
p[0]=(float)random(100)/100.;
p[1]=(float)random(100)/100.;
p[2]=(float)random(100)/100.;
p[3]=(float)random(100)/100.;
printf("p[0]=%f p[1]=%f p[2]=%f p[3]= %f\n\n", p[0], p[1], p[2], p[3]);
S0 = 3-p[1]-p[2]-p[3];
S1 = 0;
for (i= 1; i <4; i++)
for (j=i+1; j < 4; j++)
S1 += (1-p[i])*(1-p[j]);
S2 = (1-p[1])*(1-p[2])*(1-p[3]);
q = p[0]*(S0 - S1 + S2);
term = p[0]*p[1]*p[2]*p[3];
term_1 = 1.0/term;
n_r = (int)term_1; term_2 = q*n_r; dup = (int)term_2;
printf (" n_r = %d dup = %d \n", n_r, dup);
}
```

We have run several times the above mentioned program and obtained the following results:

p[0]	p[1]	p[2]	p[3]	n_r	dup
0.82	0.93	0.64	0.29	7	4
0.85	0.66	0.22	0.33	24	19
0.38	0.71	0.24	0.56	27	9
0.6	0.44	0.14	0.95	28	15
0.13	0.74	0.24	0.54	80	9
0.95	0.47	0.48	0.4	116	109
0.04	0.92	0.64	0.22	192	6
0.21	0.44	0.48	0.07	322	66
0.12	0.12	0.37	0.16	1173	139
0.52	0.24	0.42	0.01	1907	990

Table 1. Number of retransmissions and duplicate error-free packets with different probabilities

In the Table-1, P(A_1) = p[0], P(A_2) = 1 - p[1], P(A_3) = 1 - p[2], P(A_3) = 1 - p[3]. The number of retransmissions N_r = n_r and dup stands for the number of duplicate error-free packets. We have shown the table sorted on n_r. We also obtained the plot illustrated in Figure 3, which shows the interrelationship between n_r and dup.

We can clearly see that as n_r increases monotonically dup doesn't show any such monotonic increasing property. This is apparent from Table 1 where we observe that for n_r = 116 dup = 109 and for n_r = 192 dup = 6.

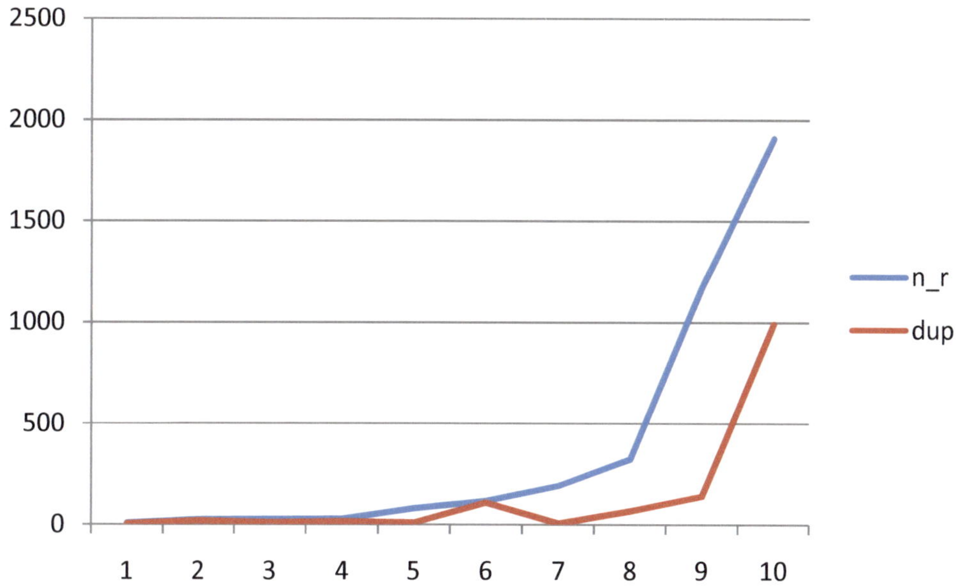

Figure 3. Illustration of the interrelationship between the number of retransmissions and duplicate error-free packets

5. Conclusion and future work

In this chapter we have analyzed and experimented with the number of retransmissions and the number of duplicate error-free packets transmitted in a generalized framework of data transmission model. As a future work we can consider some distribution on the probability of delayed transmission as a function of time.

Author details

Pinaki Mitra

Address all correspondence to: pinaki@iitg.ernet.in

Department of CSE, IIT Guwahati, Guwahati, India

References

[1] William Stallings, *Data and Computer Communications*, 8th Edition, Pearson.

[2] Prakash C. Gupta, *Data Communications and Network*, Prentice Hall India.

[3] Ajit Pal, *Data Communications and Computer Networks,* Prentice Hall India.

[4] Dimitri Bertsekas and Robert Gallager, *Data Networks,* 2nd Edition, Prentice Hall India.

[5] Lin, S., Costello, D., and Miller, M. "Automatic-Repeat-Request Error-Control Schemes." *IEEE Communication Magazine,* December 1984.

Chapter 5

Weaving Complex Patterns — From Weaving Looms to Weaving Machines

Stana Kovačević and Ivana Schwarz

Additional information is available at the end of the chapter

http://dx.doi.org/10.5772/61091

Abstract

A chronological overview of the historical facts of the art of hand weaving, with an emphasis on the development of large and complex patterns is provided. The development of looms or weaving machines through history for weaving fabric with large patterns is described. Hand weaving complex patterns from primitive weaving on frames to more complex weaving with two weavers operating one loom was especially explored. The beginning of Jacquard weaving and its application even to the electronic Jacquard looms has been emphasized. An analysis of the development of patterned fabrics with multiple effects was carried out. The analysis was conducted according to patterns using the shortened version of drawing a sample. The results show that it is possible to determine the technique of weaving and construction parameters of each fabric pattern that vary in color and/or in a weave used. The most accurate way to highlight the contours of the points of the transition of one effect into another, namely by shifting warp and weft interlacing points on the edges of each pattern (opposing thread float), is described. Based on the obtained results and finding an identical yarn, it is possible to replicate the pattern on the Jacquard loom.

Keywords: Hand weaving, looms, Jacquard weaving machine, complex patterns, fabric analysis

1. Introduction

The term textile is usually associated with a cloth that is produced on a loom or weaving machine as a fabric manufactured by interlacing warp and weft threads. Weaving is one of the oldest crafts and has a long historical development. Studying the history of the production of fabrics is based on archaeological finds, pictorial representations, frescoes, stone monuments, archival documents, etc. One of the valuable records is the Bible, which mentions the intricate network of Solomon's Temple. In Greek myths, Arachne, the weaver is mentioned who was turned into a spider by Athena; Homer describes decorative veils of the Temple of Athena. The beginnings of fabric production date back far into the past and cannot be precisely defined. The oldest sites as evidence of fabric production are certainly records and drawings in China and Egypt, carved in stone or clay pottery, which date back to 12,000 BC. It is assumed that the first interlacing of warp and weft occurred much earlier in making dwellings from brushwood. To create a roof area and better insulation of the space, mud and leaves were used, which were coiled and intertwined with each other in two directions (warp and weft directions). Thus, a solid and long-lasting roof structure was made. After that the manufacture of baskets, bags, and other household and transport articles developed. This interweaving of brushwood was probably the forerunner in the development of fabric making. By hand twisting twigs, leaves, and animal hair, an initially coarse, and afterward a finer thread was made from which fabric was manufactured by interlacing [1]. Despite coarse weaving, these fabrics had a definite advantage when worn compared to fur, especially in the summer. The preserved drawings show that fur was the first to have been used to cover the body. Fabric has gradually assumed the function of the fur and was primarily used to protect the body from bad weather. Sensitivity of textile products, and the tendency to decay, represents a particular difficulty in assessing the beginnings of making fabrics so that drawings of looms carved in stone or clay are more reliable and older evidence of the beginnings of the craft of weaving [2].

2. Production and description of fabrics with large and complex patterns

Weaving has always represented an extremely complex operation that requires not only a certain knowledge of the technological process, but also a way of creating patterns on the fabric. The production of yarns from different fibers and its transformation into a fabric is one of the oldest human crafts and technologies. Various historical textiles and materials, as an invaluable part of cultural material heritage, testify to social background and time of origin [3]. Throughout history more complex fabrics, especially silk fabrics with gold or silver threads, represented a status symbol of high society and were of great value. Most often they were used for garments, church vestments, and furniture decoration. Old simple fabrics without or with smaller patterns made from wool, linen, cotton, or similar fibers, are also greatly appreciated and carefully preserved [4].

The manufacture of closely woven silk fabric with multicolored Jacquard patterns always required a weaver with certain technological knowledge, skill of transferring a pattern on the fabric, and sense of matching colors in the pattern. Woven fabric is formed by interlacing warp and weft threads according to a weave and pattern in order to produce a compact fabric with an appropriate design. In addition, the appearance of the fabric depends on the density of warp and weft threads, composition and fineness of yarn, thread tension, loom type, and the weaver

http://dx.doi.org/10.5772/61091

(precision, patience, skill, knowledge of weaves, matching of density and color, etc.), especially when it comes to hand weaving.

It is assumed that wool is the oldest fiber that was used for making fabric. However, the oldest preserved samples are not from wool, although records mention wool processing. Egyptians wrapped mummies mainly in linen, which did not decompose, and in addition, their religion did not allow taking animal fibers into the tomb; this is the main reason why wool fabrics are not the oldest preserved samples. Over time, certain skills and weaving techniques preserved in secrecy in narrower communities were developed. Increasingly more valuable and more beautiful fabrics recognizable by regions and communities of origin began to appear. The value of a fabric was assessed according to the raw materials' complexity of manufacturing and pattern size. The fabric was evaluated according to the raw material (silk), fineness, density, and the complexity of manufacturing, had mostly a high price, and in some regions it was used as currency. Valuable preserved patterns belonged to the higher class, for vestments and appropriate clothing for various ceremonies.

Sensitivity of textile products and tendency to decompose represent a particular difficulty in assessing the beginnings of fabric manufacturing as well as manufacturing complexity. The oldest preserved samples of linen fabrics were taken from Egyptian mummies and date back to 5000 BC, when woven fabrics of certain dimensions were also woven. Most of the preserved fabrics with complex patterns were woven from natural silk. China had a centuries-old monopoly in the production and processing of natural silk. However, this monopoly was threatened as early as in the 12th century by the Italian city of Luca, which was the most famous producer of silk in the West. In the 13th century, the Silk Guild of Paris produced famous green silk fabrics woven with motifs of birds, lilies, and vine leaves. After that, the German city of Regensburg was known for its half-silk fabrics and Venice for samples of brocade with metallic golden threads. The following are some familiar names of fabrics woven on handlooms and their description [5].

Velvet is a fabric produced by weaving weft and warp threads in the form of loops, which are then cut to create the pile effect on the fabric face. It may be warp effect if warp threads are cut or weft effect if weft threads are cut. In hand weaving velvet is mostly made with patterns, and the pattern creates the pile effect which can be cut in different heights and in combination with loops (as terry cloth) depending on the pattern or the weaver's imagination. Fabrics with greater pile densities are called velours.

Brocade is a heavy, Jacquard-type fabric with a raised pattern or floral design. Traditionally, the pattern is produced with gold or silver weft threads. Weft threads are actually made from linen or wool wrapped with thin flat metal which is gold-plated or silver-plated.

Damask fabric is a firm and glossy Jacquard-patterned fabric made of silk, wool, linen, cotton, or synthetic fibers, with a pattern formed by weaving. It is characterized by the combination of satin and sateen weave, made with one warp and one weft in which, generally, the satin warp and the sateen weft weaves interchange. The figures or the designs are in the weft and the background is in the warp. Twill or other binding weaves may sometimes be introduced.

The term originally referred to ornamental silk fabrics, which were elaborately woven in colors, sometimes with the addition of gold and other metallic threads. Damask weaves are commonly produced today in silk, linen, or linen-type fabrics that feature woven patterns featuring flowers, fruit, forms of animal life, and other types of ornament.

Double cloth is a kind of woven fabric woven from two sets of warp threads and two sets of weft threads. Through weaving, two distinct fabrics are made, connected by the binding weft or warp threads. The fabric is very strong and compact and was mostly woven from wool in hand weaving.

Gros de tours fabric is a ribbed silk fabric made with a two- or three-ply warp interlaced with organzine and tram weft. Rep weave creates longitudinal or transversal stripes on the fabric. Stripes can be more or less distinct. This weave type is defined as plain weave with double threads in one thread system (mostly double weft threads) forming a ribbed appearance of the fabric. The fabric is made with a high density so it is necessary to draw into more than two heddle shafts with straight or broken draft. This disburdens shafts and warps are arranged in several shafts although they bind in the same way, so they do not cover in the fabric which is possible, especially if they are drawn into the same dent of the reed. Thus, the fabric appearance is nicer and more uniform because each thread has a certain width within the fabric and they are mutually parallel in spite of a high density. In this manner double threads that bind in the same way create stripes.

Lame is a shimmering fabric that is created by combining metallic threads, often of gold or silver, and natural or synthetic threads into a woven fabric.

Lampas is a figured textile in which a pattern composed of weft floats bound by a binding warp is added to a ground fabric formed by a main warp and a main weft. The ground fabric is mainly woven in plain, twill, or satin weave or their derived weaves with the smallest weave units. The weft, which creates the pattern, may be in the same color as the main weft or the decorative (multicolor) one, binding in plain, twill, or satin or only floats on the fabric face as extra weft-figured and/or brocaded.

Lisere is a fabric with a pattern created by the floating main weft.

Taffeta is a dense plain woven fabric. In hand weaving these fabrics are mainly woven figured with lance (extra) or brocaded wefts.

Extra weft fabric – the weft creating the pattern on the fabric is inserted over the entire fabric width. It is on the fabric face only where the pattern is created.

Brocaded weft fabric – the weft creating the pattern is on the face of the fabric and is seen only where the pattern is, and it is not woven in the other parts of the fabric. For this reason the weft consumption is lower which is especially significant if gold or silver threads are used. During insertion the brocaded weft is not cut for each weft thread, but at the end of the pattern it is returned into the pattern in the next shed, and thus relatively strong edges of the pattern are made. Pattern design of fabrics with brocaded wefts is performed with the fabric face pointing downward which facilitates and accelerates the weaving.

http://dx.doi.org/10.5772/61091

3. First looms

At the beginning of weaving longitudinal threads (warp threads) were hung next to each other on a bar with hanging weights forming one set of threads which were then interlaced with horizontally arranged threads (weft threads). Wefts were inserted, beaten up from bottom to top. This way a fabric of specific structure, softness, breathability, and comfort was produced.

In the old records weaving frames are described. The warp was taut during weaving, previously dimensioned fabrics were woven and later they were joined laterally and transformed into simple articles of clothing.

The form of the first loom changed through history and it cannot be said for certain whether the first loom was vertical or horizontal (Fig. 1). Vertical looms take up less space and are more suitable for weaving outside dwellings, but their height required prolonged abnormal body posture and standing work, especially in the beginnings of weaving when the warp was hung to weights and the weft was inserted from top to bottom. This made weft beating up more difficult. No higher warp and weft densities could be achieved, warp tension was irregular, and each weft was inserted by interlacing the warp with fingers, until the formation of the shed was invented. Vertical looms were developed mostly for weaving loosely woven fabrics in the warp direction, but with very high strength. The assumption is that the tapestries, rugs and carpets were mainly woven on vertical looms with larger widths where the colored weft conceals the loosely woven warp and creates a pattern. By tightening the warp on both sides of the frames, thereby eliminating the weights for tightening the warp, it is possible to weave from bottom to top, which made weft beating up easier and achieving a higher weft density. Additional bars and strings enabled to form two sheds, specifically for weaving in plain weave; this increased the fabric production and quality of these looms. Vertical looms are more suitable than horizontal ones for weaving larger patterns, according to the picture, which can be easily placed under the warp (no shafts that prevent placing the picture below the warp as on horizontal looms) as well as drawing pictures or drawings on the warp before weaving. This has resulted in unique items of immeasurable values. Despite slower work, and thus lower productivity in relation to the horizontal loom, vertical looms were probably more numerous at the beginning of weaving. Because of their simpler construction, vertical looms allowed achieving the utilization of the whole warp whereby it is possible to weave fabrics with selvedges on all four sides to make one article of clothing without cutting. This fact is confirmed by the preserved samples of fabrics (e.g., oldest woollen fabric woven in Europe with the edges on all four sides found in Bosnia and Herzegovina 1983; fabric age dates back to 3550-3800) [6]. It can be interpreted from the loom drawings carved in stone or clay that they are more reliable and older evidence of the beginnings of the craft of weaving and weaving process than the preserved samples of fabrics [7].

Horizontal looms are more suitable for weaving in dwellings; weavers wove in a sitting or squatting position. With the introduction of shafts and treadles weaving on the horizontal loom became easier because of the use of the feet (to press treadles) and hands (weft beating). Horizontal looms experienced a noticeable development as weaving the most complex patterns. Their productivity rose, with the introduction of the shuttle with a cop, especially in

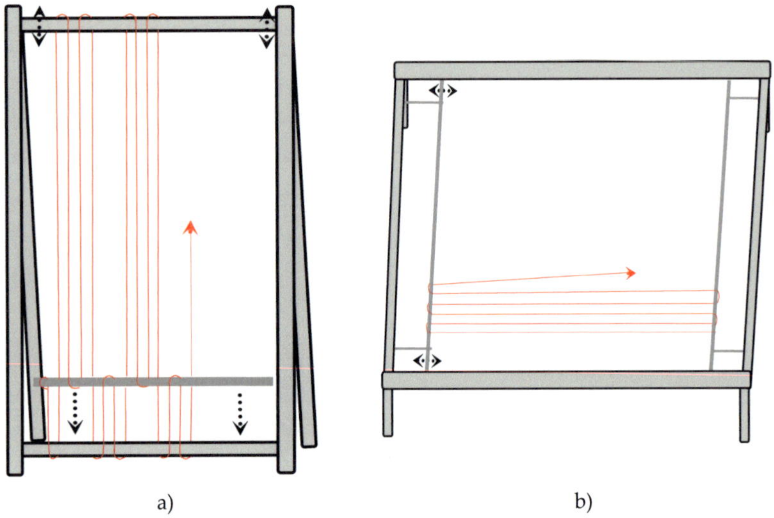

Figure 1. Warping on the loom in the form of a frame to produce a fabric with selvedges on all four sides; a) vertical loom, b) horizontal loom

case of simpler patterns. Their development was in several directions, e.g., with regard to raw materials, weaving processes, and pattern complexity. Over time, the increasing weavers' skills became more apparent in interlacing warp and weft threads creating more complex weaves, using different raw materials, finer yarn counts, and greater thread densities. In weaving greater lengths and widths of fabrics were achieved, and fabrics were later cut and sewn into more complex items of clothing. Larger and more complex patterns requiring an excellent skill and knowledge of weavers were created. The yarn was dyed with natural dyestuffs and woven, or less frequently, the fabric was dyed after weaving. Woollen fabrics were sometimes filled in order to produce warmer and softer articles of clothing. At the same time looms were adapted for weaving more complex patterns and lightweight fabrics. In this regard, horizontal looms were at an advantage: the fabric had regular and uniform dimensions across the width; it was possible to create a shed by pressing the treadles with a variety of connecting shafts and treadles, which made it easier to insert the weft; easier and faster weaving; better quality fabrics; and more possibilities of pattern design. Horizontal looms became increasingly complicated when making larger patterns and larger dimensions with extreme fineness and density of threads, which required a complex and longer warp and weft preparation. For larger fabric widths looms with two pairs of treadles and with two weavers working in pairs were constructed (Fig. 2). Sometimes, weavers were also artists or worked with them closely in weaving large patterns. Art paintings, which corresponded to proportions of drawings and colors, were adapted and woven. This required long training of weavers for weaving such patterns and reading the records of the pattern, but also artistic talent in matching colors according to the model, and skill in using different weaving techniques and knowledge of weaves that were the most suitable for the development of certain patterns and designs. Horizontal looms were specialized in weaving specific raw materials (for lightweight silk fabrics, heavy woollen fabrics, etc.), in the technique of weaving (velvet, terry, damask, lampas, etc.), in the complexity of a pattern (with extra floating wefts (lance), with brocaded wefts, with metal (gold-plated or silver-plated wefts etc.). The complexity of making patterns begins with the use of fiber (e.g., natural silk, linen, wool, cotton), yarn count (from the finest silk threads to coarse wool and metallic threads), and warp and weft density amounting to 150 threads/cm.

On the first looms, the warp had a relatively small length which sufficed for making an article of clothing. That the skill of weaving existed in the Stone Age is witnessed by findings of bone needles and dressmaking equipment made of stone. One of the more valuable artifacts are stone weights for the warp, which originate from Troy 2500 years BC, confirming the use of looms at that time [8].

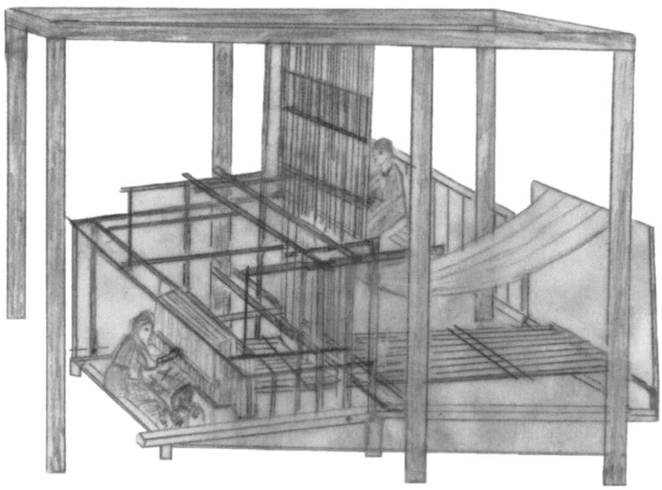

Figure 2. Horizontal loom for larger fabric widths

4. Development of looms and weaving machines for more complex weaving patterns

The first looms for making larger patterned fabrics with larger repeats that were predecessors to the Jacquard loom had a very complex construction (Fig. 3a). Before making a Jacquard loom with programming cards, making patterns with a great weave unit in combination of different weaves required a very complex weaving preparation and weaving itself [9]. A stencil plate (Fig. 3b) was prepared according to a painting, which was usually an artist's work of art. A copy of the painting and sometimes the original work was divided into squares that were 1 cm wide and 1 cm long.

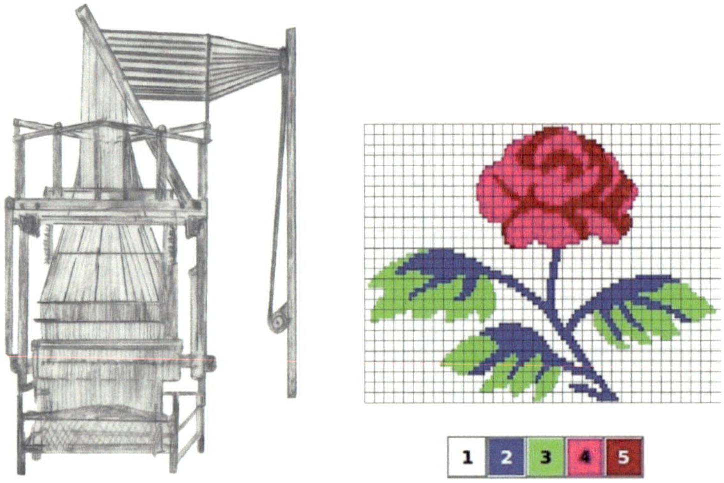

a) b)

Figure 3. a) Weaving loom for making old complex patterns before the Jacquard weaving loom, b) picture prepared for weaving with number of effects in the picture

First, it was necessary to determine the density of ends/1 cm and to calculate the total number of warp threads (density of ends/1 cm multiplied by the fabric width to be woven with the addition of threads for selvedges). According to the painting, the number of effects on the fabric pattern, which were distinguished by color and / or weave, were determined. One thread system was single-colored (or hidden) and another multicolored (creating the fabric face). In such patterns, the warp is usually single-colored, and the weft is in different colors, which creates a pattern on the fabric face. In this way, the preparation of single-colored warp is simpler and faster. According to this weaving technique, the warp threads were hidden within the fabric by beating up the weft, whereas the multicolored weft created versatile effects on the fabric. In order to hide the warp threads on the fabric face, the warp threads should be loosely warped, tauter, stronger, twisted, and preferably thinner than the weft threads. This relationship is mostly based on experience, now known as double weft, triple weft, or multiweft fabrics. In multiwarp fabrics the weft is invisible, the warp threads create a pattern, and they are finer and denser than the weft threads. Warp threads were dyed before warping, so warp preparation and warping lasted longer, but weaving was faster because the weft was mostly single-colored. A group of cords for drawing warp threads for the corresponding effect is assigned to each effect on the picture or the fabric. One of the weavers pulled the cords (weaving with two weavers) according to the corresponding effect over the whole warp width. The other weaver had to watch out for the effect in the direction of the fabric width and insert the corresponding weft color. In order to insert the warp thread over the whole fabric width, the cords for each effect in the fabric fell had to be drawn. The weft color corresponding to the effect was inserted in that fabric segment where this effect was drawn according to the pattern and the program for raising the cords for this effect depended on the weave. After the weft thread had been inserted over the whole fabric width for all effects being woven at that moment, the weft threads were beaten up, and the process was repeated for the next weft

http://dx.doi.org/10.5772/61091

insertion. The pattern was woven face downward for easier weft insertion, and a mirror was used to control the fabric pattern. The number of picks/cm depended on weft thickness and beat-up force. The thinner the weft thread and the stronger the beat up force of the weft, the more beautiful the pattern, with more distinct and accurate pattern contours created, but weaving time was longer or more weft threads had to be inserted per 1 cm. Similarly, the density of warp threads affected the quality and appearance of the fabric pattern. The greater the warp density and the finer the warp threads, the more distinct were the contours and the fabric pattern was more beautiful and more faithful. Fabric patterns made by hand weaving large patterns from natural silk, especially in combination with gold-plated and silver-plated threads (brocades) are of exceptional value [10].

The first weaving loom on which punched cards raised and lowered warp threads according to a program was invented by Joseph Marie Jacquard [9] in 1805 (Fig. 4). This loom raised warp threads with the help of a program card. This eliminated pulling cords according to effects. One cord was pulled attached to the prism and moved the card for one weft thread. By pressing the treadle the needles would come into contact with the card. The card is read for each weft and usage of hooks and the harness raises warp threads (punched hole in the card) or remains in the lower shed (unpatched hole in the card) over the whole width.

Figure 4. Jacquard loom (1805)

A punched hole in the card means that the needle passes through it and pulls the hook because it is engaged with the knife and will raise all the cords fastened to this hook and the heddles hung on the corresponding cords, and thus the threads drawn into the corresponding heddles. If there is no hole in the card, the needle remains on the card surface and does not pull the hook, it is not in contact with the knife and will not pull it, as cords or heddles and warp threads will remain in the lower shed position. One or more weft threads are inserted by colors and fineness, depending on fabric effects in this fabric segment. The technique of weft insertion by effects is the same as on the previous loom without punched card, only the shed is inactive until all weft threads are inserted as required by the pattern. This weaving technique using punched card (Jacquard punched card loom) achieves higher production, enables easier operation, and machine operation is built for one weaver and there are fewer possible faults [11].

5. Development of automatic Jacquard looms

Some inventors attempted to build a mechanically driven machine back in the eighteenth century. Cartwright applied steam power to drive the loom. Mechanical machines increasingly dominated in weaving mills of that time. Industrial production began in the second half of the eighteenth century. In 1894, Northrop invented a pirn changer, a mechanical device that provides automatic supply of weft in looms by changing the contents of the shuttle without stopping the loom. Large-patterned fabrics became possible because of weft pattern forming (two shuttle boxes on each side of the machine or 3 multicolored weft threads) and the use of the Jacquard power loom (Fig. 5).

Figure 5. Mechanical Jacquard shuttle loom with weave design

http://dx.doi.org/10.5772/61091

The production increased a number of times; the machines were improved dramatically and new manufacturers appeared. Weft pattern forming was made possible by adding four shuttle boxes on each side of the machine, which allowed up to 7 multicolored weft threads. Automatic shuttle looms were improved and used for nearly one century and played a predominant role in the historical development of weaving. Shuttle-less weaving machines appeared in the mid-twentieth century (Fig. 6) [12].

The advantage of shuttle-less weaving machines is their high productivity, higher product quality, and easier and more comfortable machine operation. On shuttle-less weaving machines, weft threads are inserted into the shed only from one side; they are unwound from the supply package outside the machine which causes problems for making strong and quality selvedges on the fabric. These machines run at high speeds because the weft inserter picks up each weft individually and it is lighter than the shuttle. On shuttle looms, the shuttle inserts the weft on the weft cop; they are slow and cannot achieve the speed of shuttle-less looms. The basic construction of shuttle-less looms does not differ significantly from the construction of shuttle looms; the only exception is the weft inserter. All devices on the loom can be classified into four basic groups: warp-unwinding devices, shed forming devices, weft inserters, and fabric winding devices.

The looms are divided into the following according to shed formation: cam loom, dobby loom, and Jacquard loom, and according to weft insertion: projectile, rapier (rigid and flexible), air jet, and water jet looms. Besides single-phase weft insertion (one weaving cycle is one weft thread), there are multiphase weaving looms. The weaving machine being developed in this direction is serial shed weaving machine which inserts 4 weft threads at each moment over the fabric width, and its productivity amounts to 6.000 m of weft/min. The other devices are weft patterning devices, warp and weft stop motions, weft accumulators, control devices, etc. [13].

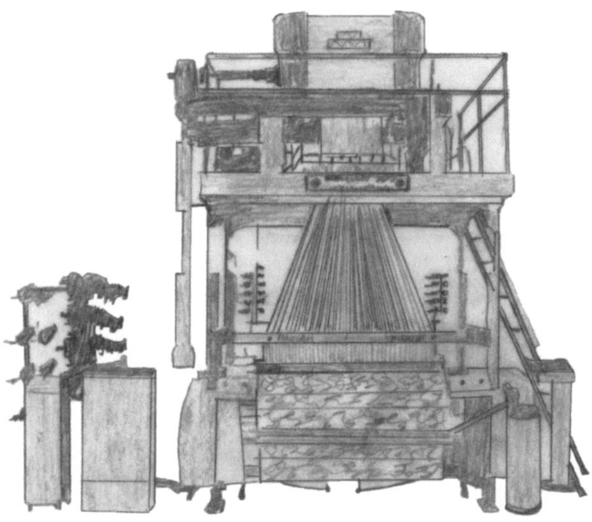

Figure 6. Electronic Jacquard shuttle-less weaving machine

Electronic Jacquard shuttle-less weaving machines have several advantages over looms with weave design. Electronic Jacquard allows better and more productive weaving. Using CAD/CAM systems in weaving shortens the pattern preparation for weaving by several times. The CAD (Computer Aided Design) system processes fabric design parameters and adapts the pattern (figure) for weaving. The CAM (Computer Aided Manufacturing) system for weaving allows controlling the production process of the loom and contains the data required for the process of weaving. CIM (Computer Integrated Manufacturing) for the process control of the whole weaving mill allows complete overview of the production, Internet contact with each loom allowing remote control, service, correction, and repair of the program.

The latest developments in Jacquard weaving are harness-less looms (Fig. 7). The upper part of the Jacquard loom (Jacquard) is "lowered" directly above the bottom machine performing the weaving process. In this way, the machine height has been considerably reduced and no harness is necessary that has to be replaced after a certain time period. These weaving machines have certain drawbacks compared with Jacquard looms with harness, but their development goes on indicating a new, more cost-effective way of weaving fabrics with large patterns.

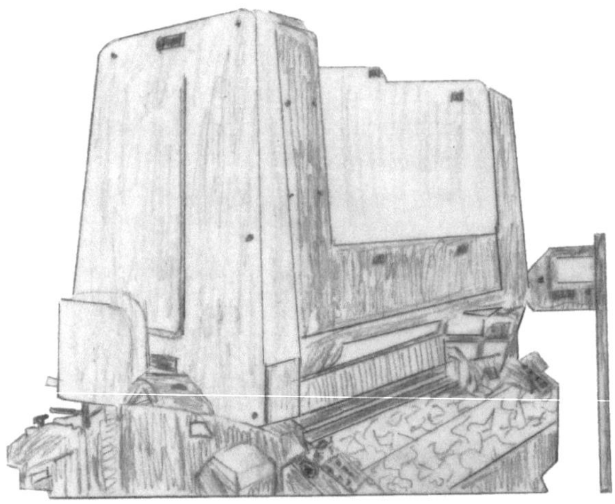

Figure 7. Electronic Jacquard loom without harness (Grosse's Unished Jacquard)

Despite the rapid development of the textile industry in the twentieth century, hand weaving has remained in handcraft industry, even in factories for manufacturing carpets having high value and quality. Today, Mexican and Persian carpets made on vertical hand looms are famous for their long life and beauty throughout the world.

6. Experimental part

The silk fabric, which will be analyzed here as an example, was woven on the Jacquard loom (Table 1, Fig. 8). The ground warp is black and has an even density across the fabric width.

The ground warp (black) and the ground weft (black) are woven in plan weave across the whole fabric surface, regardless of whether it is the ground fabric or the pattern. Various weaves and color combinations of the warp and weft compose fabric effects. The weaving machine is equipped with two warp beams: one warp beam is intended for the ground warp, and the other warp beam is intended for the pattern warp.

Warp density (ends/cm)	Ground warp (black)	92
	Pattern warp (in 7 colors)	50
Weft density (picks/cm)	Ground weft (black)	35
	Pattern warp (blue and red)	36
Fineness of warp threads (dtex)	Ground warp (black)	44
	Pattern warp (in 7 colors)	94
Fineness of weft threads (dtex)	Ground weft (black)	82
	Pattern warp (blue and red)	242

Table 1. Basic parameters of the tested fabric

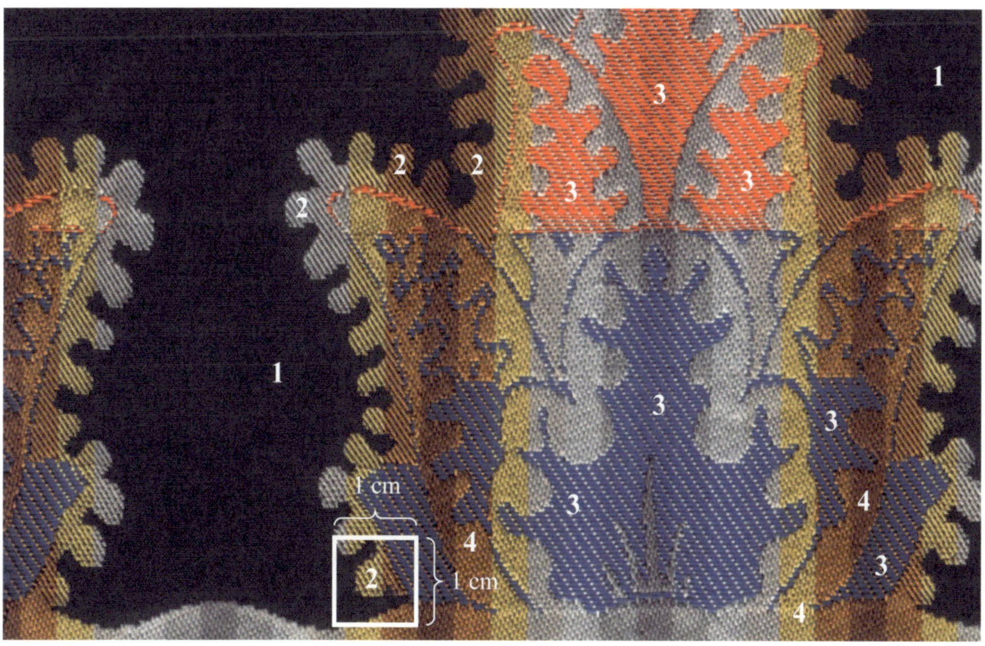

Figure 8. Fabric pattern with a multicolored warp (ground: black, pattern: cream-colored, pale gray, gray, dark gray, yellow, brown, and dark brown) and multicolored wefts (ground: black, extra floating weft: blue and red) 1-4 effects created by the pattern

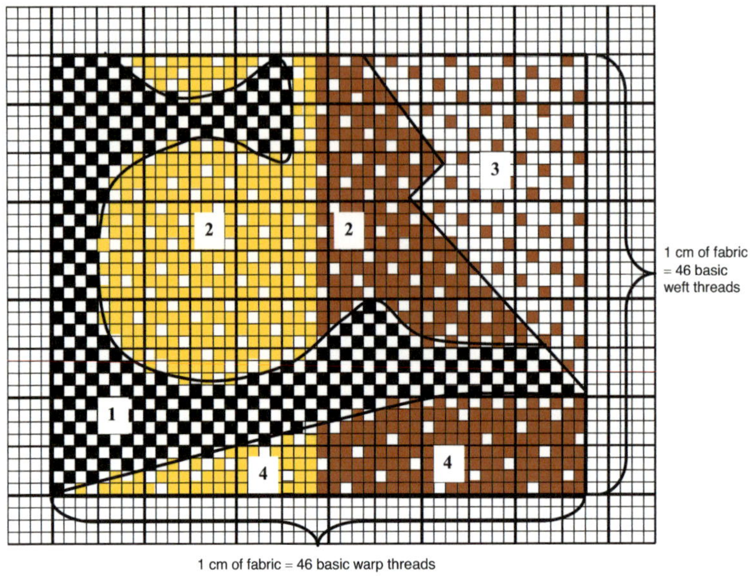

Figure 9. Weave design for 1x1 cm fabric (pronounced square on the fabric pattern, Fig. 8)

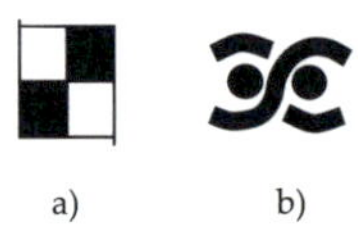

Figure 10. The first effect or ground fabric is created by the warp (black) and weft (black) in plain weave; a) weave unit; b) cross section in the weft direction

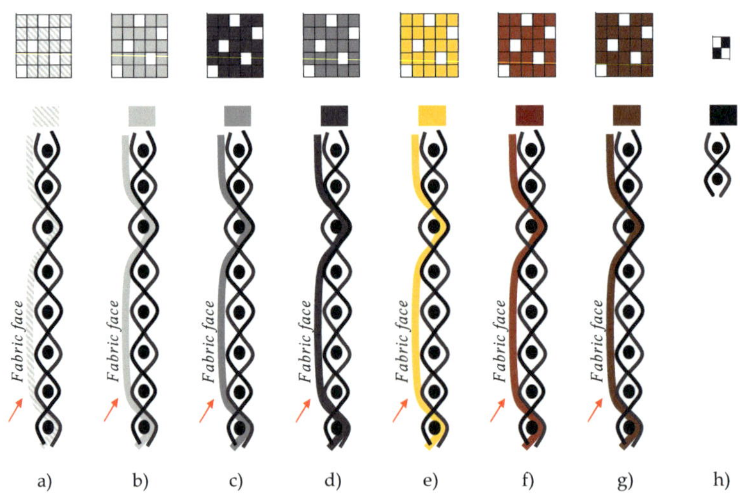

Figure 11. The third effect is created by different colors of warp threads; a) cream-colored, b) pale gray, c) gray, d) dark gray, d) yellow, f) brown, g) dark brown; they interlace with the ground in 5-end warp satin; on the back side, ground warp (black) and ground weft (black) interlace in plain weave

http://dx.doi.org/10.5772/61091

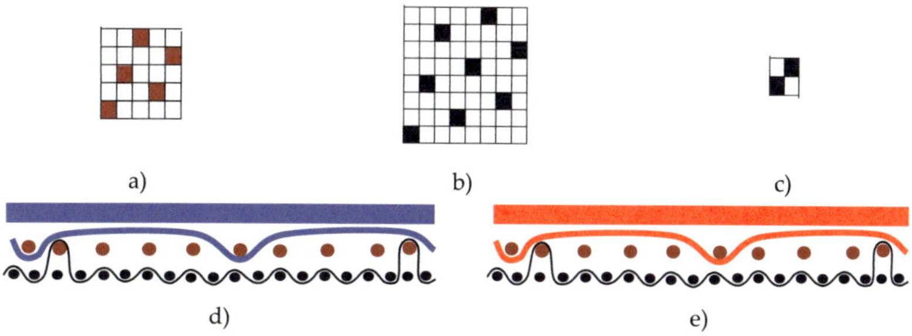

Figure 12. The third effect is created by the red and blue weft threads which interlace with pattern warp threads in 5-end weft satin; on the back side, the ground warp (black) and ground weft (black) interlace in plain weave, in the middle section of the fabric the pattern warp threads interlace with the ground warp threads in 8-end warp satin; a) 5-end weft satin, b) 8-end weft satin; c) plain weave, d) cross section with blue weft on the fabric face, e) cross section with red warp thread on the fabric face

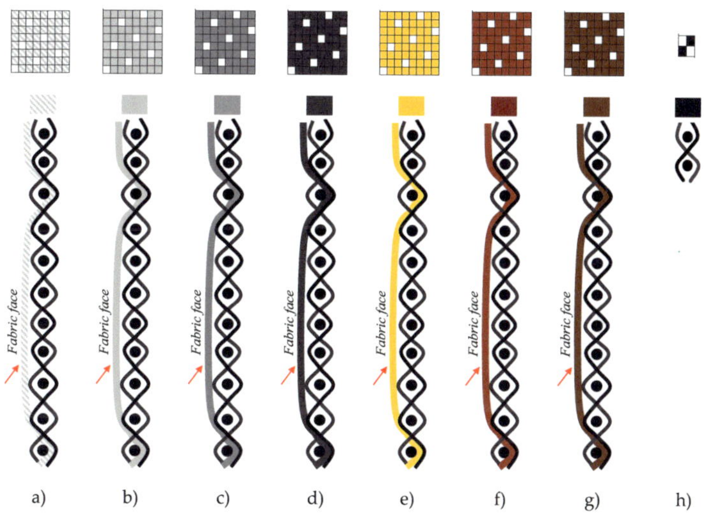

Figure 13. The fourth effect is created by different colors of the warp threads: a) cream-colored, b) pale gray, c) gray, d) dark gray, e) yellow, f) brown, g) dark brown interlace with the ground weft in 8-end warp satin; on the fabric face, the ground warp (black) and ground weft (black) interlace in plain weave

7. Discussion

Marked effects on the fabric are only those created by different weaves. Effects by the colors of warp and weft threads are not specifically identified, but they are visible on the fabric and in the weave development. In addition to the ground weft, two extra floating wefts (lance) (red and blue) creating the pattern with the pattern warp run in the weft direction. The pattern warp and lance weft threads are on the back side of the fabric and do not interlace in the ground fabric part. The fabric was woven on the Jacquard weaving machine, and the pattern does not include the weave unit. The weave is drawn on the appropriate graph paper which corresponds to the ratio of warp-to-weft density. Since the weave unit is relatively large, the total number of warp and weft threads is not covered by the analysis of drawing the weave. That is why the weave of only one fabric part has been drawn, covering all the effects. It is important that in the transition from one into another effect there is opposing thread (warp or weft) float. In that case, a certain displacement of interlacing points in favor opposing thread float is allowed which was done in the figure representing the weave development (Fig. 9).

The looms, which were used to weave pattern fabrics before Jacquard weaving, had to be adjusted for this kind of interlacement which required a specific skill and knowledge of the weaver. In transition areas of effects the weave is disrupted, meaning that the weft interlacing point is applied in one effect opposite the warp interlacing point in the other effect and vice versa in order to point out the boundary between two effects. In hand weaving, the weaver had to pull weft threads past the shed in transition areas of effects in order to create opposing thread float. Pulling the weft thread through in these areas was difficult and resulted in faults in the weave, breaks of warp threads and longer weaving. This problem disappeared when Jacquard looms were introduced because cards were punched according to the prepared weaving pattern where opposing thread float had been drawn. In this way, the shed was opened for each warp with opposing thread float in areas between effects, resulting in a great increase of productivity and fabric quality. Punching cards for each warp thread allowed forming one shed for all effects and weft colors to be inserted into the corresponding shed. Thus, it was unnecessary that the other weaver raised warp threads according to the effects because the card was punched for all warp threads over the whole fabric width or all effects across the fabric width at this moment.

Each effect is explained in detail below:

- The first effect is the ground part of the fabric, where the ground warp (black) and the ground weft (black) are interlaced in plain weave (Fig. 10).

- The pattern warp in different colors (cream-colored, pale gray, gray, dark gray, yellow, brown, and dark brown) and the ground weft makes the second effect (black). The weave on the fabric face (pattern warp and ground weft) is 5-end warp satin (A 4/1 with a leap to the right), and plain weave on the backside (ground black warp and ground black weft) (Fig. 11).

- The third effect is made by the lance weft (black or red) and the pattern warp in different colors (cream-colored, pale gray, gray, dark gray, yellow, brown, or dark brown) which interlace in weft 5-end satin (A ¼ with a leap 3 to the right), and the weft is visible on the fabric face. In one fabric segment the red weft and the blue weft interchange, and the logic rule of their interchange in the pattern is not visible. In the second effect, the ground weft and the pattern warp in different colors interlace in 8-end warp satin (A 7/1 with a leap 3 to the right) which is the middle layer of the fabric, while the ground warp and the ground weft interlace in plain weave on the backside of the fabric (Fig. 12).

http://dx.doi.org/10.5772/61091

- The fourth effect is made by the pattern warps in different colors (cream-colored, pale gray, gray, dark gray, yellow, brown, and dark brown) and the ground weft (black) in 8-end warp satin (A 7/1 with a leap 3 to the right) (Fig. 13).

8. Conclusion

The historical development of weaving dates back before the Neolithic period. Interweaving brushwood in building dwellings is the forerunner of interlacing warp and weft threads in fabric making. At first, a coarse textile fabric with low-density simple weaves was made; by increasing yarn fineness and improving weaving procedures, the fabric became finer and more comfortable. Gradually, weaving skills and techniques kept in secrecy in the narrower community were developed. Larger and more complex patterns were introduced. Weaving looms changed throughout history and were adapted to types of fabrics woven on the loom. The manufacture of fabrics with larger and more complex patterns, before Jacquard looms had been constructed, was performed on more complex designed Jacquard looms operated by two weavers according to a drawing or a picture. Productivity was very low regardless of the skill and training of weavers. Yarn preparation (spinning, dyeing, warping, drawing in) required an operator's knowledge and skills. With the advent of the Jacquard loom in the early nineteenth century, productivity rose several times as well as the quality of weaving; one weaver operated the loom.

The principle of the shed formation using a card with punched and unpunched holes did not change for decades. Only the Jacquard gauge changed and the possibility of weaving with several differently interlacing warp threads increased. With the advent of shuttle weft insertion power looms productivity increased several times. Machine operation was simpler, because of the automatic pirn change in the shuttle, one weaver could operate several looms, and the fabric became better and cheaper. The electronic Jacquard loom with CAD/CAM weaving system increased the pattern preparation for weaving as well as the possibility of making larger patterns or greater weave units. Moreover, the consumption of cards was eliminated as well as trial weaving, and greater faults in weaving were not possible.

In order to analyze the pattern of the fabric, it is necessary to define the face of the fabric, the warp and weft direction, and the size of the weaving unit if the fabric pattern involves the weave unit. The analysis of the large pattern fabric involves analyzing the fabric construction parameters for each design or the development of each fabric effect. Due to the complexity of making the weave in the pattern, the analysis is often conducted only on a small area of fabric (1×1cm), preferably that it encompasses all designs or analyzes each design separately. The implementation of such an analysis is important because in order to prepare a certain amount of yarn by color, warp length, and the method of winding on one or two warp beams. Besides, it is important to define the density, contraction, and yarn fineness. According to warp thread density, fabric width, and arrangement of colors in the fabric width it is possible to calculate the total number of warp threads as well as the number of threads by colors. According to the weaves per designs, it is necessary to draw in warp threads into the harness according to the weave, design program, and lowering warp threads (manual pulling the hooks before the Jacquard, punching the cards after the Jacquard or pattern preparation using the CAD/CAM weaving system). In order for the fabric pattern to have precise contours (or boundaries) between individual designs, it is necessary to subsequently move the warp and weft interlacing points (to achieve opposing thread float as much as possible) on the edges of the design.

Author details

Stana Kovačević and Ivana Schwarz*

*Address all correspondence to: ivana.schwarz@ttf.hr

University of Zagreb, Faculty of Textile Technology, Zagreb, Croatia

References

[1] Harris J. (ed.) *5000 Years of Textiles*. London: British Museum Press in association with The Whitworth Art Gallery and The Victoria and Albert Museum; 2004.

[2] Kovačević S. *Ručno tkanje*. Zagreb; 2003.

[3] Schwarz I., Kovačević S. Hand weaving – tradition of the future. In: Salonen R., Heikkila P. (eds.) Proceedings of 7th AUTEX Textile Conference: From Emerging Innovations to Global Business, Tampere University of Technology, Finland; 2007. p. 1121-1128.

[4] Johnstone P. *High Fashion in the Church: The Place of Church Vestments in the History of Art, from the Ninth to the Nineteenth Century*. Leeds, UK; 2002.

[5] Markowsky B. Europäische seidengewebe. Köln, Kunstgewerbemuseum der stadt Köln, Des 13-18. Jahrhunderts;1976.

[6] Bender Jørgensen L., Grömer K. Textile Remains from Early Bronze Age Tumulus grave No. 16 in Pustopolje. Project CinBA, 2010-2013, Creativity and Craft Production in Middle and Late Bronze Age Europe, Project Leader: Joanna Sofaer, University of Southampton; 2011.

[7] Car G., Meder F. Conservation report on the conservation of the fragments of the cryptic mantle from prehistoric tumulus no. 16. Zagreb: Pustopolje Kupres owned by the Franciscan Museum and Gorica Gallery, Livno, Bosnia and Herzegovina, Croatian Conservation Institute; 2010.

[8] Gillow J., Sentence B., editor. Paul Haupt V. *Atlas der Textilien*. Bern, Stuttgart, Wien; 1999.

[9] Joseph Marie Jacquard. http://en.wikipedia.org/wiki/Joseph_Marie_Jacquard (accessed 25 February 2013)

[10] Pavičić S. Tekstil, Paramenta + Crkveni tekstil Hrvatskog povijesnog muzeja. Zagreb: *Hrvatski Povjesni Muzej*; 1998.

[11] Adanur S. *Handbook of Weaving*. Florida, US; 2000.

[12] Barlow A. *The History and Principles of Weaving by Hand and by Power*. UK; 2007.

[13] Kovačević S., Dimitrovski K., Hađina J. *Procesi tkanja*. Zagreb; 2008.